Practice Game Theory

*Get a Competitive Edge in Strategic Decision-Making,
Avoid Getting Outplayed, and Maximize Your Gains.*

By Albert Rutherford
www.albertrutherford.com

Copyright © Albert Rutherford 2022. All rights reserved.

No part of this publication may be reproduced, stored in a retrieval system, or transmitted in any form or by any means, electronic, mechanical, photocopying, recording, scanning or otherwise, except as permitted under Section 107 or 108 of the 1976 United States Copyright Act, without the prior written permission of the author.

Limit of Liability/ Disclaimer of Warranty: The author makes no representations or warranties regarding the accuracy or completeness of the contents of this work and specifically disclaims all warranties, including without limitation warranties of fitness for a particular purpose. No warranty may be created or extended by sales or promotional materials. The advice and recipes contained herein may not be suitable for everyone. This work is sold with the understanding that the author is not engaged in rendering medical, legal or other professional advice or services. If professional assistance is required, the services of a competent professional person should be sought. The author shall not be liable for damages arising herefrom. The fact that an individual, organization of website is referred to in this work as a citation and/or potential source of further information does not mean that the author endorses the information the individual, organization to the website may provide or recommendations they/it may make. Further, readers should be aware that Internet websites listed in this work might have changed or disappeared between when this work was written and when it is read.

For general information on the products and services or to obtain technical support, please contact the author.

I have a gift for you…

Thank you for choosing my book, Practice Game Theory! I would like to show my appreciation for the trust you gave me by giving The Art of Asking Powerful Questions – in the World of Systems to you!

In this booklet you will learn:
   -what bounded rationality is,
   -how to distinguish event- and behavior-level analysis,
   -how to find optimal leverage points,
   -and how to ask powerful questions using a systems thinking perspective.

Visit www.albertrutherford.com to get your FREE GIFT: The Art of Asking Powerful Questions in the World of Systems

Table of Contents

PRACTICE GAME THEORY	3
I HAVE A GIFT FOR YOU…	7
TABLE OF CONTENTS	9
INTRODUCTION	13
Normal Form and Extensive Form Games	14
Cooperative and Non-cooperative Games	19
Symmetric and Asymmetric Games	20
Simultaneous and Sequential Games	21
Constant Sum, Zero-Sum Games, and Non-Zero-Sum Games	22
CHAPTER 1: THE NASH EQUILIBRIUM	25
Pure-Strategy Nash Equilibrium	28
Mixed-Strategy Nash Equilibrium	33
Pure- and Mixed-Strategy Nash Equilibria	37
CHAPTER 2: WOULD YOU CHALLENGE A MONOPOLY?	43
Subgame Perfect Equilibrium	43
Health care problems	52

CHAPTER 3: THE ESCALATION GAME   57

Backward Induction   57

CHAPTER 4: SHOULD I STAY OR SHOULD I GO?   64

Multiple Subgame Perfect Equilibria   64

The Mixed-Strategy Algorithm   68

How do we calculate payoffs?   78

Exercise 1: The Battle of the Sexes calculation.   84

Exercise 2: Health care problems.   84

CHAPTER 5: GAME STACKING   87

The Purpose of Punishment   92

CHAPTER 6: CREDIBILITY AND NON-CREDIBILITY   99

Non-Credibility   107

Many steps, one conclusion   110

How to find the subgame Perfect Equilibrium?   113

The Problem   120

SOLUTION FOR THE HEALTH CARE PROBLEMS EXERCISE   123

BEFORE YOU GO…   135

REFERENCE   137

ENDNOTES   145

Introduction

Game theory is the study of strategic decision-making; a framework for using mathematical models to understand the behavior and motivation of competing, rational players.[i] Game theory was coined in 1940 by John von Neumann and Oscar Morgenstern. Its importance has grown ever since. A good metric supporting this statement is that since the 1970s, twelve leading economists won the Nobel Prize in Economic Sciences for their contributions in game theory.[ii] It is present in multiple fields where logic is the governing factor: business, finance, economics, politics, sociology, and psychology. It provides strategic thinking patterns, moves, and explanations to help you develop exceptional decision-making skills.

The games you find in game theory are mathematical tools. We need three things to define these games:

- "the players of the game,
- the information and actions available to each player at each decision point,
- and the payoffs for each outcome."[iii]

There are multiple game types in game theory, each serving a purpose and helping analyze a problem. Based on how many players are in the game, their strategy profile (cooperation, symmetry), the nature of the game (simultaneous, zero-sum), we distinguish these game types:

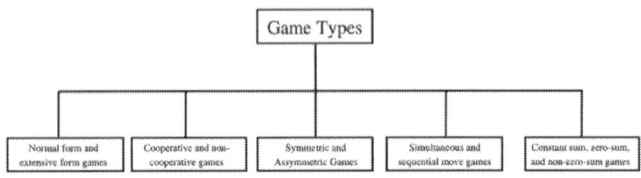

Picture 1: Game Types in Game Theory

Normal Form and Extensive Form Games

Normal-form games are illustrated as tables or matrices. The number of rows equals the number of Player 1's strategies. The number of columns is the number of Player 2's strategies. The matrix, therefore, shows all the potential outcomes based on every possible strategy a player could use.[iv]

Jim ↓ / Tom →	Left	Right
Up	4; 2	-2; -2
Down	0; 0	2; 4

The matrix above is a two-player game. The actions of Jim are presented horizontally, and Tom's actions are shown vertically. Each player has two strategies. The number of rows and columns illustrates this. The payoffs are the numbers in each box. The first number is the payoff of Jim. Meanwhile, the second number is the payoff of Tom. If Jim chooses the strategy to play Up and Tom plays Left, then Jim receives a payoff of 4, and Tom gets 2. When a game is illustrated in normal form, we expect each player to act simultaneously. Or, at the very least, without knowing the moves of the other.

If players have information about the moves of other players, we are usually talking about a *sequential* game that uses the extensive form. Extensive form games are illustrated with game trees. It's a diagram that shows that decisions are made at different points in time. The nodes show when a choice was made. The payoffs are shown at the bottom of each branch. Game trees help players predict all decisions and counter-decisions in any game. We use *backward induction* to solve these games. [v]

Extensive form games can have perfect or imperfect information. In a game with perfect information, all players know all previous moves of other players. When Jim moves, he knows where he is in the game and also knows about Tom's previous decisions. Chess is an example of an extensive form game with perfect information.

In a game with imperfect information, some information about the moves is missing. Examples of imperfect information games are poker and bridge.[vi]

Extensive form games can be categorized as games with complete or incomplete information. In a game with complete information, the players know the game's structure. What does this mean? The players know the order in which they are supposed to move. They know every possible move in each position and the payoffs for every outcome. Games in the real world are usually not like this; they don't have complete information. We assume complete information in basic game theory modelling as games with incomplete information are harder to analyze.[vii]

A perfect and complete extensive-form game is presented on a game tree, including the following aspects:

- the players;
- all the move opportunities of all players;
- what players can do at their every move;

- what every player knows about every move;
- the payoffs gained for every combination of moves by each player.

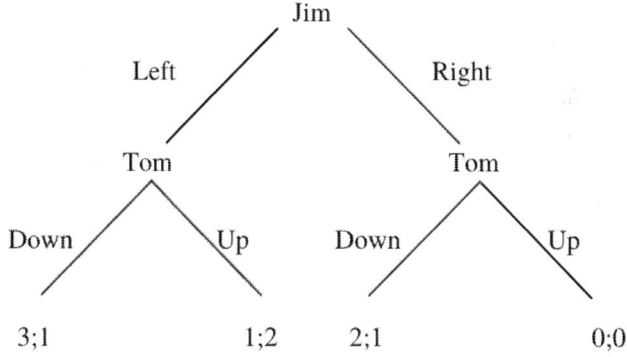

Picture 2: Complete extensive form game with perfect information.

The game above has two players, Jim and Tom. Jim's choices are to go Left or Right. Tom will go Down or Up based on what Jim decides. The numbers at the end of the branches show the payoffs to the players. Let's look at the two numbers on the far left of the game tree, (3;1). 3, in this case, is Player 1's, Jim's, payoff. 1 is Player 2's, Tom's, payoff. The labels beside each line (Left, Right,

Down, Up) are the name of the action that line represents.

Jim starts this game, the first node illustrates his first move. The game flows the following way: Jim chooses between Left and Right. Tom observes Jim's choice and then chooses between Down and Up. Their payoffs are written at the bottom of the tree. This game has four outcomes shown by the four terminal nodes of the tree: (L,D), (L,U), (R,D) and (R,U) and the outcomes of these results are (3;1), (1;2), (2;1), and (0;0).

We can further notice that if Jim plays Left, it's in Tom's best interest to move Up as 2 (Tom's outcome if he chooses Up when Jim moves Left) is bigger than 1 (Tom's outcome if he chooses Down when Jim moves Left). However, if Jim moves Right, Tom will choose Down, as 1 is greater than 0.

When a game tree is illustrated as in Picture 2, there is clarity about which player goes first and what they choose. There are games – with imperfect information – with no clarity about what Jim chose. This is the case with simultaneous games or where there is hidden information. When this situation occurs, the player bound to move next (in Picture 3 below, this player is Tom) can't differentiate between the two nodes. He doesn't know which was the decision of Jim – left or right. We can see this represented by the dotted lines. We will learn how to solve such games later in this book.

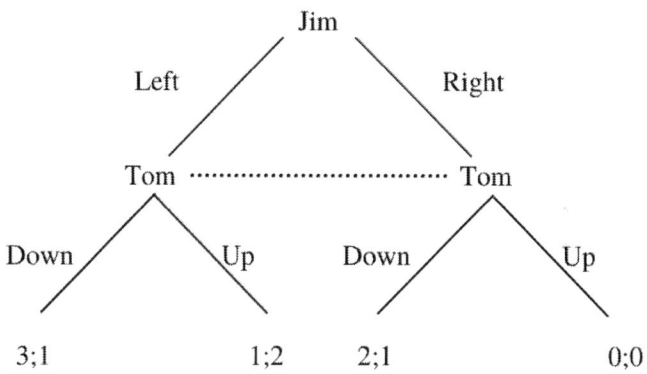

Picture 3: Extensive Form Game with Imperfect Information.

Finally, sometimes, a player doesn't know the payoffs of the game, how many opponents they have, or the type of their opponents. These games have *incomplete information,* and are otherwise known as Harsányi transformation. They bring into the picture the idea of "nature's or God's choice." [viii] In this book, we will not discuss games with incomplete information.

Cooperative and Non-cooperative Games

Some games encourage players to form alliances. What matters is that the allies commit to having obligations to one another.[ix] Game theory's primary purpose is to predict which players will form an alliance, what joint action they will take, and what gains they can expect. Coalitions, merges, oligopolies, and trusts can be examples in the real world.

Non-cooperative games aim to predict individual players' actions and payoffs. In non-cooperative games, participants either can't form alliances due to the nature of the game, or the agreements they make are based on credible threats that they need to self-enforce. Players make individual choices to maximize their payoff.[x]

Symmetric and Asymmetric Games

In symmetric games, the payoff of a strategy depends only on the other strategies employed, not on who is playing them.[xi] If players can be replaced without changing the gains or losses in the strategy, the game is symmetric. All players adopt the same strategy. The Prisoner's Dilemma and Stag Hunt are two classic examples of this. We will learn about both later.

Asymmetric games have different strategies for each player. The Ultimatum Game is one such game. This was a popular economic experiment proposed by the Hungarian-American John Harsányi in 1961.[xii] In this game, one player, the proposer, gets x amount of money. Then, the proposer needs to split it with another player, called the responder. Once the proposer chooses the amount and shares the news with the responder, they may accept it or reject it. If the responder accepts, the money is divided per the proposer's wishes. If the responder rejects, both get nothing. The players know in advance what happens if the responder accepts or declines.[xiii]

Some games can have the same strategy for both players and still be asymmetric.

Simultaneous and Sequential Games

In simultaneous games, players either make their move at the same time or, if they don't move at the same time, they are blind to the other players' moves. The card game War is an excellent example of simultaneous games as it has players simultaneously showing their top card, with the highest card winning. The goal of the game is for one player to win all fifty-two cards.

In sequential (or dynamic) games, you know about previous actions of other players. Chess is a sequential game. The information a player has is often not perfect or complete. It may, in fact, be very little knowledge. Think about the game Battleship. It is a two-player game on paper with ruled grids where each player marks a set number of "battleships" of various sizes. The locations of the battleships are hidden from the other player. Each player sequentially calls their "shots," trying to "sink" the other person's battleships as the game flows. The winner is the player who destroys the other's fleet first.

Another difference between simultaneous and sequential games is that simultaneous games often use normal form, while sequential games use extensive form.

Constant Sum, Zero-Sum Games, and Non-Zero-Sum Games

A constant sum game is when the sum of the outcome for all players remains constant even if their outcomes are different. The constant is an unchanging resource that player decisions won't influence. Depending on the strategies used, each player gets more or less of the available resource.

A zero-sum game is a type of constant sum game where the total benefit is shared in various ways among all players and always adds to zero. One player gains at the equal loss of other players.[xiv] Imagine a large pepperoni pizza. Depending on who slices it, one diner might get that weirdly small slice that can be eaten in two bites while another diner gets a slice as big as their head. The pizza size is the same, but the second diner got more than their fair share. Matching pennies is a classic zero-sum game in game theory, and we will learn about it later. Chess and go are the best-known zero-sum games in the board game world.

In non-zero-sum games, like the Prisoner's Dilemma, the outcome has net results bigger or smaller than zero; one player's gain does not necessarily equal the loss of other players.

Chapter 1: The Nash Equilibrium

The Nash Equilibrium's function is to predict the outcome of the players' decision-making and stands for all players' *strategy profile* in a game. Nash Equilibria are stable -yet not always efficient – strategies where no player can benefit by unilaterally changing their strategy. "If a unique Nash Equilibrium exists for the game, then all players are expected to converge to the state represented by the equilibrium if they are all rational – that is, each player aims to choose the strategy that maximizes its utility function."[xv] Players have control over their choices. In other words, the Nash Equilibrium acts like a rule that no player wants to break. [xvi] Some games have more than one Nash Equilibrium, some have none, and some have only one. The most illustrative and simple example for a Nash Equilibrium is the Prisoner's Dilemma.

The Prisoner's Dilemma[xvii]

The game goes as follows, two criminals, Jim and Tom, have been picked up by the police to be interrogated on suspicion of committing a major crime. The police has no evidence, but they are certain of Jim's and Tom's guilt. The two criminals are led into separate interrogation rooms by police officers who offer each one a deal. They are told that they must serve one year each on a lighter charge if both stay silent. This is not the best deal but could be worse, is it? But then the chief of police enters the room, and as a part of a well-rehearsed police strategy, gives them each a better deal. If Jim rats out Tom, he gets immunity for his testimony. But Tom will serve ten years in prison, now for a provable crime. Tom gets the same offer. If both talk, they will both serve eight years having confessed to their crimes. Should either criminal talk or remain silent? Jim and Tom must decide without knowing the decision of the other.

Jim ↓ / Tom →	Stays Silent	Betrays
Stays Silent	-1; -1	-10; 0
Betrays	0; -10	-8; -8

The table above shows the rules of the game. The numbers are the number of years spent in prison. Mathematicians call these tables a game matrix. What will Jim and Tom do?

The answer is obvious, they both should stay silent. They serve a year at the taxpayers' expense, and that's that. But there is no loyalty among thieves. And Jim wouldn't swear on Tom's trustworthiness. Jim's internal monologue looks like this:

"I have only two choices: to speak or not to speak. If Tom and I both stay silent, we both serve one year. But ... but if Tom opts for the Fifth and I talk, I walk! I should talk," ponders Jim. "However, if Tom is a rat and I don't say a thing, I will rot in jail for ten years! If I talk, at least I will only serve eight." And then slowly it sinks in to Jim: *"Whatever choice Tom makes, I am better off if I talk either way!"*

This thought process sums up the logic of the classic Prisoner's Dilemma game. Both Jim and Tom, being rational players who want to maximize their benefit, will conclude that they should talk so they will both get eight years of jail time. The (-8; -8) strategy is a Nash Equilibrium because if either player kept silent, they would have gotten ten years instead. Thus, neither was incentivized to change strategies. The game's goal is not winning but to make a decision that neither player will regret. If Jim kept silent and Tom spoke, Jim would have regretted his choice as he had to go to prison for ten years instead of eight, and vice versa. Therefore, the stay silent/betray strategy pair is not a Nash Equilibrium.

Pure-Strategy Nash Equilibrium

A pure-strategy Nash Equilibrium is a set of strategies for each player, where no player has an incentive to change their strategy given what other players are doing, as their payoff wouldn't increase by the change. Players don't randomize between two or more strategies.[xviii] Both players are definitely choosing one strategy or definitely choosing another strategy. Thus, their strategic choice is deliberate. The Stag Hunt is a great example of the pure-strategy Nash Equilibrium.

The Stag Hunt[xix]

Jim and Tom decide to go out to hunt. Other hunters have informed them that there are two hares and one stag in the nearby forest. They each need to decide what to hunt. A stag is harder to capture than hares, so they need to work together. The matrix for the stag hunt looks like this:

Jim ↓ / Tom →	**Stag**	**Hare**
Stag	4; 4	0; 2
Hare	2; 0	1; 1

In this matrix, we see that one stag yields eight units of meat. One hare provides only one unit. If both hunters hunt the stag, it will result in each gaining four units of meat. On the opposite end, if both hunters go after the hares, each will get one unit of food.

If Jim chooses to hunt the stag alone, he is unsuccessful while Tom snags the hares. The opposite scenario plays out the same way. If Tom goes after the stag and Jim after the hares, Tom will get zero and Jim two. These aren't Nash equilibria because in these scenarios, one player can switch strategies and get a better outcome. The player who would get zero hunting a stag can choose to hunt for hare instead and get one. The player hunting for hares can choose to hunt for a stag and get four.

This game doesn't have *strictly dominant strategies*, meaning one strategy is always better than others. Why? Let's say Jim knows that Tom will hunt a stag. Jim, in this scenario, should also go for the stag because 4 is more than 2. (See the bolded and underlined options for Jim.)

Jim ↓ / Tom →	Stag
Stag	**4**; 4

Hare	**2**; 0

But if Jim knows that Tom will hunt for hares, he will choose to hunt for hare as well—1 is more than 0.

Jim ↓ / Tom →	Hare
Stag	**0**; 2
Hare	**1**; 1

We can conclude that Jim's optimal strategy depends on what Tom is doing. This conclusion stands true in Tom's case, too. He will hunt a hare if Jim's hunting a hare and go for the stag if Jim is also hunting for the stag. Neither strategy is *strictly dominated*.

What's the solution for this game? The Nash Equilibrium is a set of strategies, one for each player, where no player has any incentive to change their strategy. Nash equilibria are stable. One player's actions are optimal given what the other player is doing. Once one player has chosen their strategy, they have no regrets about it. They can't do better if they change their strategy retrospectively.[xx]

Let's discover how to find the Nash equilibria (there are two) in the Stag Hunt game. We will do this by looking at one outcome at a time and seeing if either player can *individually do better* if they change their strategies. We will begin by looking at the outcome of the Stag-Stag scenario.

Stag-Stag (4; 4)

Jim ↓ / Tom →	**Stag**
Stag	4; 4

Would either Jim or Tom have a better outcome if they choose a different strategy? Jim would not want to change his strategy because if he went for hare, he would only get two units of meat instead of four. The same stands true in Tom's case. If he went for hare, he would also end up with two units of meat. We can conclude that the Stag-Stag outcome is a Nash Equilibrium. No one will have bitter feelings following this strategy. Except the stag. This is the best outcome for the hunters, yielding the highest reward. But as I dropped the spoiler earlier, there are two Nash equilibria in this game. Let's find the other one.

Stag-Hare (0; 2) or Hare-Stag (2; 0)

Is Jim or Tom individually hunting for a stag a Nash Equilibrium? Is this a stable strategy? No, it is not. Why? Look at the individual outcomes. If Jim knows that Tom will hunt the stag, it makes sense for Jim to change his strategy and go for the stag as four is better than two. Let's approach it from Tom's perspective. If Tom knows that Jim will hunt for hare, he will also go for hare because one is better than zero.

The opposite scenario is also true. It's important to notice that there is an individual deviation that leaves each player better off.

Hare-Hare (1; 1)

If both Jim and Tom choose the hare scenario, would they have a profitable deviation if they changed strategies? No, they wouldn't. If Jim decided to go for a stag instead, he would be getting zero instead of one. The same is true in Tom's case. If he went for the stag, he would also gain zero instead of one.

Jim ↓ / Tom →	**Hare**
Hare	1; 1

Collectively (meaning in both players' cases), this outcome is also a stable Nash Equilibrium. This is a less obvious Nash Equilibrium because, in this case, both players are worse off than if they both hunted the stag. But if there was a special reason for one of the hunters not to hunt stags, this can be a good alternative strategy to remember. If both Jim and Tom follow it, they fare somewhat better than if they didn't cooperate. It's an inefficient choice, but *Nash equilibria are not always efficient*. They are meant to be *stable*.[xxi]

Mixed-Strategy Nash Equilibrium

The Nash theorem states there has to be at least one Nash Equilibrium in all games with a finite number of moves.[xxii] Some games have no pure strategies, so there must be another strategy. We call these mixed strategies. A mixed-strategy Nash Equilibrium is a strategy "profile with the property that no single player can obtain a higher expected payoff (utility) according to the player's preference over all such lotteries."[xxiii] In other words, mixed strategies are probability distributions over one or more pure strategies. This means that the players will select their choice randomly, in equilibrium. If the mixtures are the best responses for all players, the set of strategies is a mixed-strategy Nash Equilibrium.[xxiv]

I know this definition might as well be in Hungarian, so let me elaborate with an example, the

Matching Pennies game – the zero-sum game I was talking about in the Introduction.

In this game, Jim and Tom both have a penny. They will reveal their penny to each other, either heads or tails up. Each time both show matching pennies of heads or tails, Jim wins $10. Each time they show mismatching pennies, Tom wins $10. We can illustrate this in a matrix as follows…

Jim ↓ / Tom →	**Heads**	**Tails**
Heads	10; -10	-10; 10
Tails	-10; 10	10; -10

As you can see in the matrix, if both Jim and Tom get either heads or tails, Jim wins $10, and Tom loses $10. If Jim shows heads and Tom shows tails or vice versa, Tom gets $10, and Jim loses $10. This is why we call this game a zero-sum game; the result of the two players in each box adds up to 0. Let's do the math: 10 + (-10) = 10-10 = 0. Or -10 + 10 = 0. Jim and Tom have diametrically opposed interests in this game. Zero-sum games are games that have a finite number of resources.

Why doesn't this game have a pure-strategy Nash Equilibrium? Because if Tom knows Jim is going to play heads or tails, he will simply play the opposite side to win $10. Let's prove this. Could the heads-heads choice be a Nash Equilibrium? No. Tom can deviate for a better outcome playing tails if he knows that Jim will play heads. If Jim plays heads and Tom plays tails, Jim will want to play tails too. If we examine each box, we will see that one player can have a better outcome if he changes strategies in each of them.

If both Jim and Tom are selecting heads or tails at random (flip the coins, in other words), they could assume that they will win about fifty percent of the time and lose fifty percent of the time. Let's assume Tom decides to flip the coin, getting heads half the time and tails half the time. Jim sticks to playing heads only. Here, Jim will only win half of the time, as well.

Jim ↓ / Tom →	**Heads (1/2)**	**Tails (1/2)**
Heads	10; -10	-10; 10

The opposite is also true. If Tom flips the coin and half of the time gets heads and half of the time tails, and Jim plays only tails, Jim will still only win half of the time.

Jim ↓ / Tom →	**Heads (1/2)**	**Tails (1/2)**

Tails	-10; 10	10; -10

It doesn't matter what Jim and Tom do—they can't change their outcome. They will each win or lose half of the time.

Jim↓/Tom→	Heads (1/2)	Tails (1/2)
Heads (1/2)	10; -10	-10; 10
Tails (1/2)	-10; 10	10; -10

There is no way that either will do better if they change their strategy. This is what a mutual best response is. Tom is best-responding to what Jim is doing. He can't secure a win or a loss, so he might just flip the coin and see what happens. The same stands true in Jim's case. Neither of them can gain more by changing the coin flipping strategy. Therefore, we found a Nash Equilibrium.

Not every game is so easy to guess as just flipping a coin. Let's give different weight to each outcome in the matching pennies game.

Jim↓/Tom→	**Heads**	**Tails**
Heads	3; -3	-2; 2
Tails	-1; 1	0; 0

Here, if both Jim and Tom flip tails, there is no consequence. If both flip heads, Jim gains 3 and Tom loses 3 (-3). If Jim plays heads and Tom plays tails, Jim loses 2 (-2) and Tom gets 2. If Jim plays tails and Tom plays heads, Jim loses 1 (-1) and Tom wins 1. This payoff matrix doesn't reveal an obvious Nash Equilibrium. Here, the two players are not indifferent to the outcome of the coin flipping. To solve games like this, we will need to use some—easy—math. It is an algorithm for mixed strategies. We will learn more about it in Chapter 4.[xxv][xxvi]

Pure- and Mixed-Strategy Nash Equilibria

There are games where there are pure and mixed Nash Equilibria. The most well-known example of such a game is the Battle of the Sexes. Jim and Kat want to go out on a date. Kat prefers attending a dance class. Jim's preference is to watch a rugby game. They couldn't agree where to go, just that they should meet at seven. However, this date happened in the '80s, before the age of cell phones, so they had no way to contact each other about where to go. They had to choose blindly. Jim and Kat have four possible outcome combinations:

1. Kat at dance class, Jim at dance class.
2. Kat at dance class, Jim at rugby game.

3. Kat at rugby game, Jim at dance class.
4. Kat at rugby game, Jim at rugby game.

Out of the four outcomes, the couple ends up together only in scenarios 1 and 4. Let's illustrate this in a game matrix.

Jim ↓ / Kat →	Dance	Rugby
Dance	1; 2	0; 0
Rugby	0; 0	2; 1

When they both go to dance, Kat is fully satisfied, so she gets a payoff of 2. Jim is at least with her despite his preference, so he gets 1. Both choosing rugby makes Jim very happy, so he gets 2 and Kat gets 1. In the outcomes dance-rugby or rugby-dance, one member of the couple is at their preferred place, but they are not together. They don't want to be alone, so they just go home, unhappy, both getting a payoff of 0.

Based on what we learned so far, we can detect two pure Nash Equilibria. Dance-dance and rugby-rugby are both pure Nash Equilibria as neither player could expect a higher payoff if they individually deviated. If, in the dance-dance scenario,

Jim chose to go to rugby instead, his payoff would be 0 instead of 1. Also, if Kat deviated in the dance-dance scenario and went to rugby, her payoff would be 0 instead of 2. So, there is no incentive for either of them to change. The same stands true in the rugby-rugby scenario. If Jim changed strategies and went to dance instead, he would get 0 instead of 2. And Kat would get 0 instead of 1, given she changed her strategy.

The Battle of the Sexes thus has two pure-strategy Nash Equilibria. Jim and Kat want to be together doing what the other partner preferred while sacrificing what they would have liked to do. So, we can see the incentive for cooperation. But we can also notice that their preference differs. Coordination, in this case, is not very clear. We can't be sure if they will end up in the dance-dance or rugby-rugby outcomes.

What can we do? This is where the mixed strategy comes into play. To discover the mixed strategy of this game, we need to *run the mixed-strategy algorithm* and *calculate each mixed-strategy outcome's payoffs*. We will discuss the algorithm and payoff calculation in Chapter 4. For now, I will give you the answers. (Spoiler alert, it will be your job later to run the mixed-strategy algorithm and calculate the payoffs for this game as a practice.) Here are the results:

Jim ↓ / Kat →	Dance 2/3	Rugby 1/3
Dance 1/3	1; 2	0; 0
Rugby 2/3	0; 0	2; 1

What does this mean? When Jim is mixing his strategy to go to dance with a probability of 1/3 and to rugby with a probability of 2/3, Kat is *indifferent* whether to go to dance or rugby. When Kat is mixing her strategy to go to dance with a probability of 2/3 and to rugby with a probability of 1/3, Jim is *indifferent* whether to go to dance or rugby.

After we do the payoff calculation, we will discover that both Jim and Kat would get a *payoff of 2/3* (this is not the same number as the 2/3 probability). This is how the mixed-strategy algorithm's payoffs would look like.

Jim ↓ / Kat →	Dance 2/3	Rugby 1/3
Dance 1/3	2/3; 2/3	0; 0
Rugby 2/3	0; 0	2/3; 2/3

We can conclude that choosing the mixed-strategy Nash Equilibrium yields a worse payoff for both players than choosing either of the pure-strategy

Nash Equilibria. The payoff of the mixed strategy for both Jim and Kat is 2/3, which is less than 1 or 2. Thus, even if Jim or Kat end up at their least preferred activity, they are still better off going with the pure-strategy Nash Equilibrium, which gives them at least 1 point. In other words, it is always smarter for Jim to just agree at home to go to the dance class. It is smarter for him to give up his preference for Kat's choice. Similarly, Kat also fares better if she agrees to meet at the rugby game. Her payoffs will be higher than going for the mixed strategy where they don't come to an agreement.[xxvii]

This concludes a quick Nash-Equilibrium bootcamp. In the next chapter, we will learn about a refinement of the Nash-Equilibrium called subgame perfect equilibrium.

Chapter 2: Would You Challenge a Monopoly?

Subgame Perfect Equilibrium

Imagine coming up with the perfect business idea. You dedicated so much time to realizing your dream product – it is relevant, it is unique, and you're sure it would sell... But there is a problem. Somebody else was just a little bit faster in developing and throwing this product into the market, and they are now in a monopoly position. What should you do? Ditch your dreams and keep on grinding in the rat race of your 9-to-5 job, or follow the "fortune favors the bold" slogan and enter the market? Great question! You will find an answer by learning about games with subgame perfect equilibria.

Whoa... subperfect what? Worry not, I will break this concept down for you right now.

What is a subgame?

Technically, a subgame is a part of a bigger game that can be analyzed as a standalone game—despite being a part of a larger game. The information and payoffs of the subgame originate from the main game.[xxviii] In other words, big games can have smaller games embedded within them, and we call these subgames.

What is subgame perfection?

When we're looking for subgame perfection, our task is to remove all noncredible threats from the game tree. Given that a subgame involves a noncredible threat, we can rule out the presence of Nash Equilibrium.

What does this mean? When we're analyzing strategies potentially involving Nash Equilibria, we will realize that not every move a player takes is a move they would normally want to take, but game circumstances force them. However, even in these conditions, we still expect players to maximize their gains and take the most reasonable step based on their circumstances. There are cases where a Nash Equilibrium is not a sensible move in extensive form games – we will see this soon in an example. Subgame perfect equilibrium thus alters the idea of Nash Equilibrium. "A Nash Equilibrium is said to be subgame perfect if and only if it is a Nash

Equilibrium in every subgame of the game."[xxix] Chapter 6 will go through a step-by-step breakdown of how to assess which Nash Equilibrium is also a subgame perfect equilibrium.

What is a subgame perfect (Nash) equilibrium?

Subgame perfect equilibrium is a subset and refinement of Nash Equilibrium we use in sequential games. It is a set of strategies where Nash equilibria are formed in all subgames of the original game. We could say that every "subgame perfect equilibrium is a Nash Equilibrium. But a Nash Equilibrium may or may not be a subgame perfect equilibrium. The key difference between subgame perfect equilibrium and Nash Equilibrium is that subgame perfect equilibrium requires credible threats. Consequently, the study of subgame perfect equilibrium is the study of credible threats."[xxx]

"Informally, this means that at any point in the game, the players' behavior from that point onward should represent a Nash Equilibrium of the continuation game (i.e., of the subgame), no matter what happened before."[xxxi] A Nash Equilibrium is subgame perfect if it stands true in *every subgame* of the main game *and* the main game itself.

Tom's Monopoly

So far, we learned that in extensive form games players take turns when they strategize instead of moving simultaneously. When we analyze games with subgame perfect equilibria, we're looking for credible threats. What does this mean?

The players in the game I'm about to present are Jim and Tom. Jim wants to open an axe throwing bar in his hometown in Oregon. While the idea is good, and many people are interested in this harmlessly aggressive pastime, there is one problem. Tom already has an axe throwing bar in town, being in a monopoly position. What should Jim do? Should he even enter the axe throwing bar market? And if he does, what should Tom's response be? Allow Jim to enter and share the market with him or declare a price war? Let's determine the best strategic thinking-based response Jim and Tom could give.

From Jim's perspective, entering the market is reasonable only if Tom *doesn't* declare a price war. If he does, then Jim's profit margins can shrink to such a level that entering the market would not be worth it. If Tom, the monopolist, declares a price war, his own profitability will suffer, too. Let's illustrate this game on a game tree. On the game tree, the first numbers are Jim's payoffs, him being Player 1, and the second numbers are Tom's payoff, being Player 2.

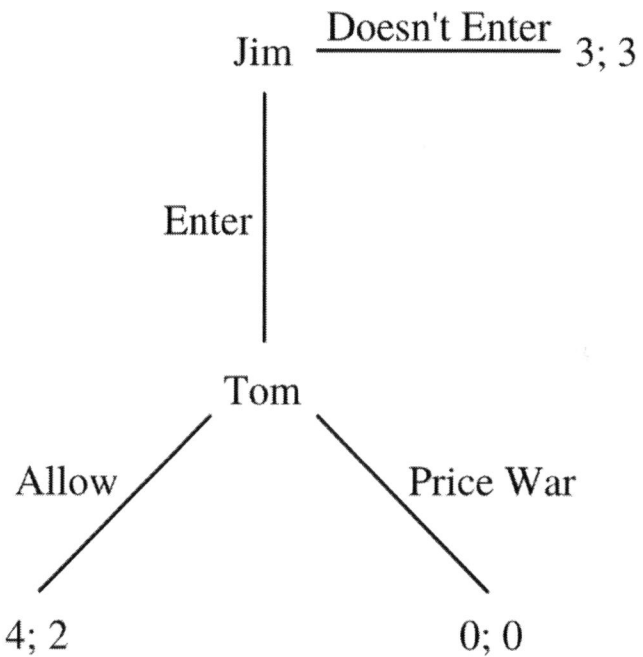

Picture 4: Monopoly Game Tree

As we can see in Picture 4, Jim can enter or not enter the market. If Jim doesn't enter, this game ends here. Tom will keep his monopoly position and get a payout of 3. Jim will keep the money he wanted to use to enter the market, so he also gets 3.

If, however, Jim enters the market, Tom will have two moves; he can either allow Jim into the axe throwing bar market or reject Jim from entering by declaring a price war. If Tom goes for the latter option, both will get 0 as the profit margins shrink into nothing. If, however, Tom allows Jim into the market, Jim will be more profitable as people will want to check out the new bar in town. Jim will get a payoff of 4, while Tom's payoff will be 2 – somewhat less than when Jim chooses not to enter the market but more than when he engages in a price war.[xxxii]

How many Nash equilibria do you think there are in this game? At first glance, you may think, two! One would be this option:

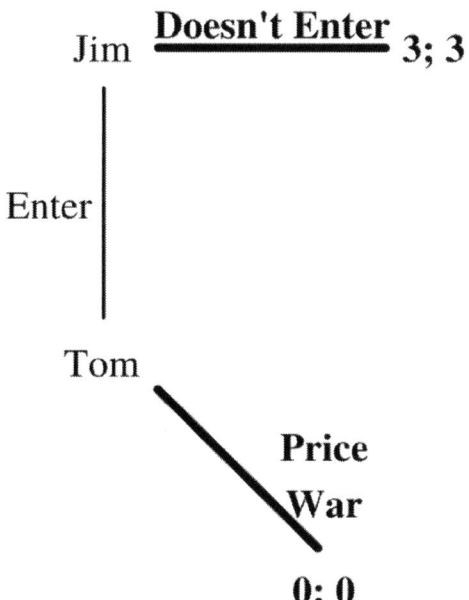

Picture 5: Doesn't Enter – Price War Scenario

Jim chooses not to enter in this scenario, meaning both gain 3. Tom's move would be to declare a price war if he had the chance – but he doesn't as the game ends with Jim's choice. If Tom could respond to Jim's move, both would fare worse, getting 0 each. Jim would never make that move though, as it wouldn't be a profitable deviation, 3 is better than 0. So, in this scenario, there is no real turn taking. As long as Jim stays out, both of them gain 3.

The other equilibrium would be this option, where Jim enters the market.

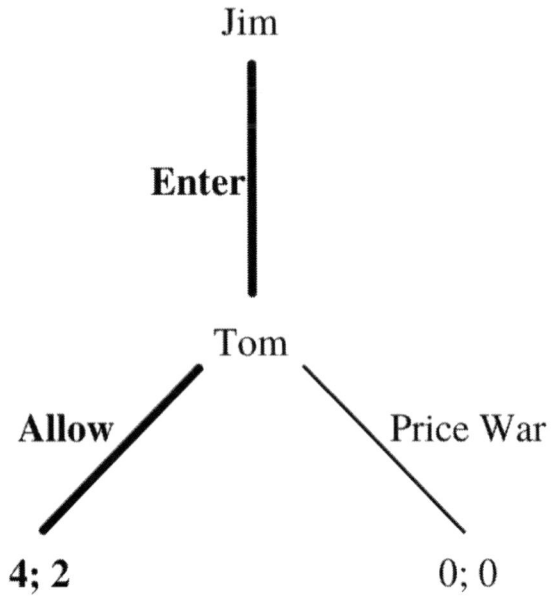

Picture 6: Enter-Allow Strategy.

Why? Because if Jim enters and Tom allows it, Jim's payoff is 4 and Tom's is 2. Why wouldn't Tom choose a price war? Because in that case, his payoff would be 0, so it is not a profitable deviation for him. Assuming Tom is a rational player (and in this book, we assume players are rational), he wouldn't choose 0 over 2.

Monopoly Tom would never want to declare war in a game where players take turns. Why? Because whenever Tom does get an opportunity to make a move, it's never in his best interest to declare a price war. Therefore, Tom would always allow Jim into the market in the scenario where Jim chooses to enter. Tom will gain no profits if he declares war.

For Jim, the only rational option to not enter the market would be if Tom threatens a price war if Jim enters. (See Picture 5 – Doesn't Enter – Price War) But Jim should be able to assess rationally that if he enters the market, it will no longer be in Tom's best interest to follow through with the price war threat. Thus, it makes no rational sense for Jim *not to enter* the axe throwing bar market. In other words, while the Doesn't Enter – Price War strategy is a Nash Equilibrium, it doesn't make sense. Tom engaging in a price war is not a credible threat.

If I illustrated this game in a chart, it would look like this, where the underlined numbers are both pure Nash equilibria.

Jim ↓ / Tom →	Allow	Price War
Enter	<u>4;2</u>	0;0
Doesn't Enter	3;3	<u>3;3</u>

As we saw, we had to cut this game into two "subgames" (Enter – Allow and Doesn't Enter – Price War). To state that this game is "perfect," we had to assess whether the threats were credible. In the they Doesn't Enter – Price War strategy, the equilibrium was maintained by *Tom's noncredible threat*. Thus, we have a Nash Equilibrium, but it is NOT subgame perfect. We identified there is a Nash Equilibrium, where Jim and Tom both maximize their benefits given the circumstances – the Enter-Allow strategy. This is a Nash Equilibrium, and it is also subgame perfect as neither players have an incentive to deviate from the Nash Equilibrium neither in the subgame, nor in the main game.[xxxiii]

When we're playing such subgames, our goal is to discard options where the threats are not realistic.

Health care problems

Let's examine another game tree with a similar structure. In this game, we don't analyze the credibility of a threat but the credibility of the commitment.

Jim lives in Europe, where they have socialized medicine. He has had health problems for a while now. The good thing about socialized medicine is that you get free health care if you're a taxpayer. The bad part about this system is that it can take months to get an appointment with a specialist. But you can always go to the ER to get a quick checkup, some pain killers and assurance that you'll survive another day.

What should Jim do? We can illustrate his situation on a game tree.

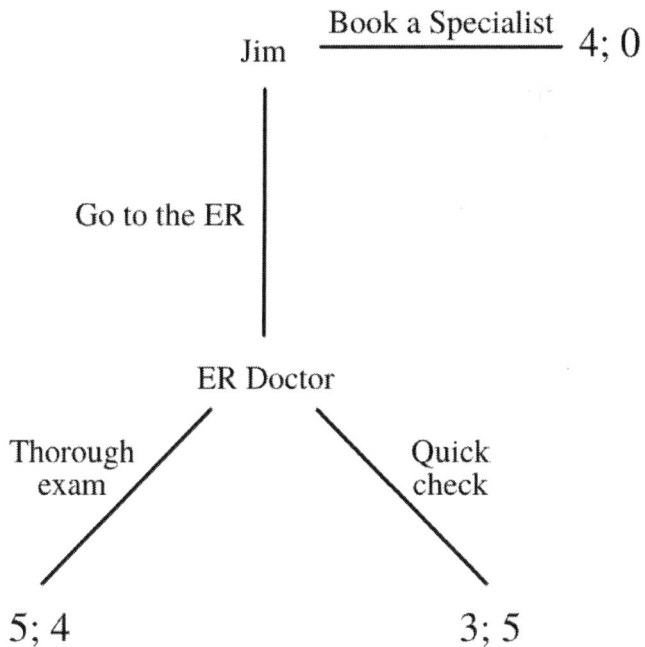

Picture 7: Jim's Medical Outcomes.

We see on this game tree that Jim has two options: going to the ER or booking a specialist. To go to a specialist could take weeks or months, so going to the ER would be definitely faster for Jim.

But the speed of the visit only helps him if the ER doctor will perform a profound health checkup, run a bunch of tests and figure out what's Jim's ailment. Thus, he would prefer a thorough examination. The ER doc, however, has a productivity-based salary, meaning, the more people she pushes through the ER, the more money she will get. (This description isn't meant to smear doctors. It is an unfortunate facet of the productivity-based medical system.) In other words, a quick check is in the ER doc's best interest. So, if Jim goes to the ER, he can expect that the doctor will perform a quick examination, write him some ibuprofen, and see to the next patient who just got his arm chopped off by an electric saw.

Knowing this, what should Jim do?

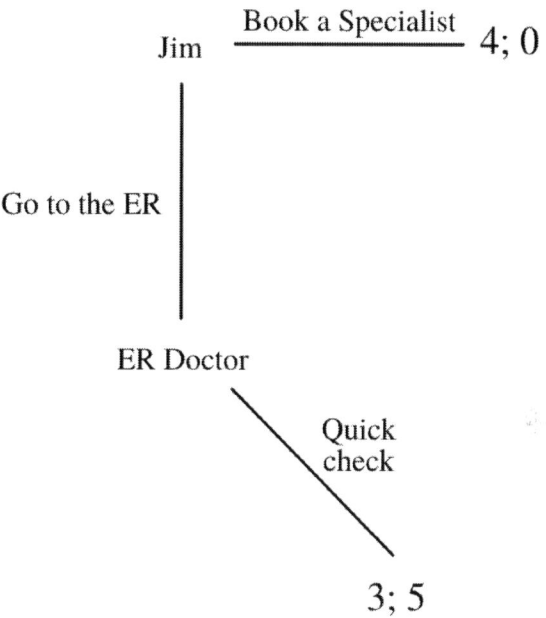

Picture 8: Jim's Optimal Outcome knowing the ER Doctor will perform a Quick check.

If we only look at this side of the game tree, clearly, Jim would do better if he just booked the specialist and waited to get a proper diagnosis. The ER doctor, in this case, would get nothing as Jim wouldn't visit the ER. Jim's payoff is higher if he waits for a specialist. So, the ER doctor won't get to make a move in this game.

Interestingly, the best outcome for both Jim and the doctor would be if they chose this strategy:

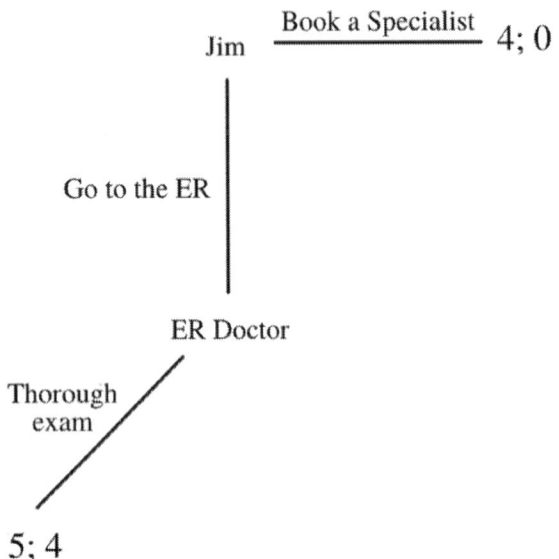

Picture 9: The strategy with the highest mutual payoffs.

If the ER doctor promised Jim that she would perform a thorough examination if Jim chose to go to the ER, both would fare better. Jim would get his diagnosis faster, having a payoff of 5, and the doctor would also get a decent payoff of 4. But the problem is, the doctor won't be able to commit convincingly. Why? Because once Jim is in the ER, the doctor's best interest is to push Jim through quickly. Or, maybe even if she would honor her word, she may get a patient with seven gunshots so she wouldn't be able to deal for long with Jim's tummy ache. Thus, both Jim and the doctor will end up where the outcome is worse for both. In this game, there is no pure strategy Nash equilibria, as each player can deviate from their choice and get a better outcome. It is also not a subgame-perfect equilibrium as the Nash Equilibrium doesn't stay true in the subgame and the main game.

Is there a mixed-strategy Nash Equilibrium? At the end of Chapter 4, you will be asked to answer this question, running the mixed-strategy algorithm and payoff calculation.

Chapter 3: The Escalation Game

Backward Induction

Backward induction is the most common Nash Equilibrium refinement for sequential games. Backward induction assumes that the players are rational, thus they stay in equilibrium, anticipating a worse outcome if they were to deviate. Players have to evaluate out-of-equilibrium moves when using backward induction. But when we assume rational players, we know there is no chance for an out-of-equilibrium play. [Binmore, 1987; Stalnaker, 1999][xxxiv]

"Backward induction is the process of analyzing a game from back to front. At each information set, we remove strategies that are dominated."[xxxv] In other words, when we use backward induction, we go to the last subgame of a finite game, analyze the strategy profile, find the best response or Nash Equilibrium, assign a payoff, and move step-by-step towards the game's beginning. Backward induction helps to prove that the restriction in a given strategy profile to any subgame is a Nash Equilibrium.[xxxvi] We follow the logic of sequential rationality, "an optimal strategy for a player should maximize that player's expected payoff, conditional on every information set at which that player has a decision."xxxvii

This sounds confusing, I'm sure, so let's look at an easy-to-follow example.

The Escalation Game

Let's assume that Jim and Tom are two kings whose kingdoms are at the brink of war. The game looks like this:

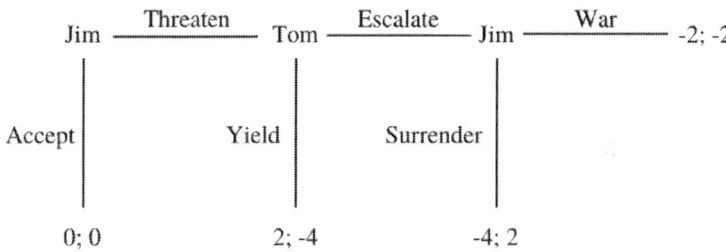

Picture 10: The Escalation Game

If Jim accepts the status quo of the two kingdoms and doesn't engage in aggression, then the game is over, and both kingdoms stay where they currently are, getting a payoff of 0 each. However, if Jim threatens Tom's kingdom, there are two responses Tom could give – he could yield to the threats, subordinating himself to Jim. This way, Jim gets a payoff of 2 (in the form of some taxes and punishments he bestows upon Tom). Tom's payoff will be -4 (exploitation from Jim).

If Tom escalates the situation and threatens Jim in return, Jim can choose one of two paths. If Jim underestimated the power of Tom, he might choose to surrender. This looks bad for Jim as he will have to pay a significant amount of money to Tom's kingdom, his payout will be -4. Tom will be on the receiving end of the bounty, getting 2. Jim could also call bluff on Tom, and decide to go to war. If they declare war on each other, both get a payoff of -2 as wars are expensive, people are killed, there is no workforce for domestic production and so on.[xxxviii]

How can we solve this game? Maybe your first instinct is to start the solving process from the beginning. But why would I talk about backward induction then? That's right. We will start at the end and work towards the beginning of the game.

Why is this a better approach? Let's put ourselves in Jim's shoes. When would be rational for Jim to start threatening Tom, overturning the status quo? It would only make sense if Jim was positive that the payoff of threatening Tom would be greater than accepting the status quo. How can he know that threatening would yield a greater payoff? By going to the end of the game and working his way backwards.

We can ask the same question in Tom's position, as well. When would he choose escalation over yielding? When he knows that escalation results in a higher payoff. The only way for Tom to know this is to check what Jim would do in the next step and work backwards.

Let's go to the end of this game and start the analysis with Jim's final decision:

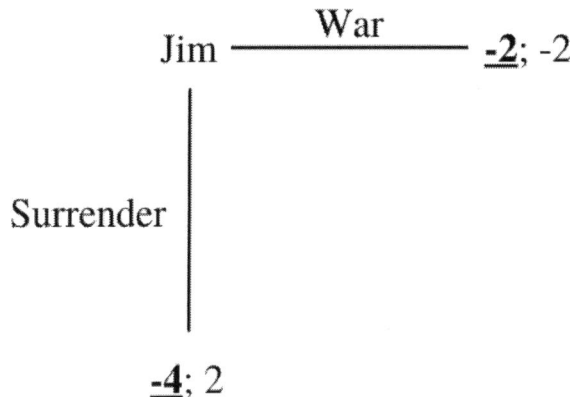

Picture 11: Jim's final decision.

Remember, this result comes from Jim threatening Tom as a first step and then Tom escalating the situation as a second step. If Jim finds himself in the scenario in Picture 11, it is better for him to declare war (-2) than surrender (-4). So, the rational decision for Jim is to go to war. Let's go back with one step and see Tom's choices.

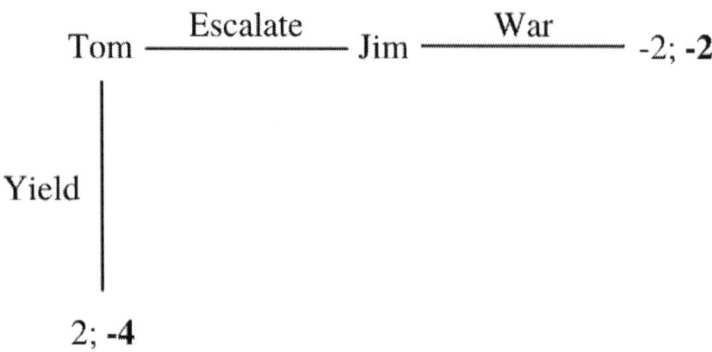

Picture 12: Tom's Choice.

Working our way back on the game tree, we can see a new game. Tom sees that if he escalates the conflict, Jim will declare war. Tom knows that Jim will not surrender as that would not be a rational choice for Jim. So, Tom's choices are: escalation with a payoff of -2 or yielding with a payoff of -4. Thus, we can conclude that if Tom is ever in this scenario, he will choose escalation as that option offers a better payoff than yielding.

Now, let's go back to the start of the game. Jim now knows that if he declares war, and Tom, knowing that Jim would declare war, chooses escalation because of Jim's initial threat, the payoff would be -2 for both players.

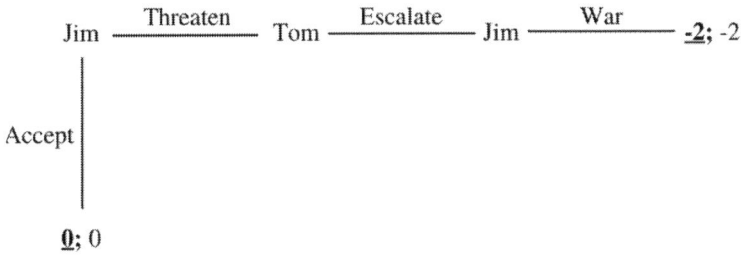

Picture 13: The full game from Jim's viewpoint.

Jim now knows that if he provokes a war, his payoff will be -2, whereas if he just accepts the status quo, his outcome will be 0. 0 is greater than -2; thus, it doesn't make sense for Jim to start a war. He chooses "Accept," and the game ends.

As long as the game is sequential, meaning each player moves after the other player, you can use backward induction and work your way backward to analyze the optimal strategy at each move point. This will lead you to the subgame perfect Nash Equilibrium. Here, it was Jim accepting and Tom escalating. But just like in the previous chapter, the game ended before Tom even got to make his move.[xxxix]

Chapter 4: Should I Stay or Should I Go?

Multiple Subgame Perfect Equilibria

We explored two games in the previous two chapters with one subgame perfect equilibrium. However, we can have games that have multiple subgame perfect equilibria.

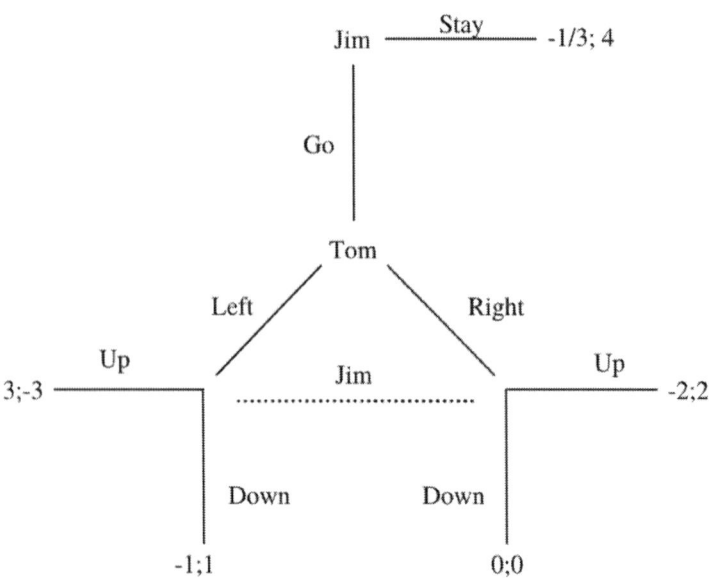

Picture 14[xl]: Game Tree for Multiple Subgame Perfect Equilibria.

64

In this game, Jim has two initial options: to stay or to go. If Jim goes, Tom can choose between left or right. And then, based on Tom's choice, Jim could select between up and down. But here is a caveat, Jim *doesn't know* where Tom went. The dotted line indicates this lack of certainty. So while we can see that Jim's responses to either scenario are up and down, he can't assess Tom's previous step. We saw this game earlier in the Introduction when I discussed games with imperfect information.

We can't use backward induction straightforwardly here as the subgame doesn't have perfect information. Jim doesn't know for sure if Tom is left or right when he chooses between up and down. Jim only knows that Tom made a move, either right or left. The payoffs further complicate Jim's life because if Tom goes left, Jim would rather go up as 3 is bigger than -1. However, if Tom goes right, Jim would choose down as 0 is greater than -2. Long story short, we can't just chop off one of the final steps of this subgame to analyze it, because we don't have enough information. So, what do we do?

We cut off the top part of this game and scrutinized the following section:

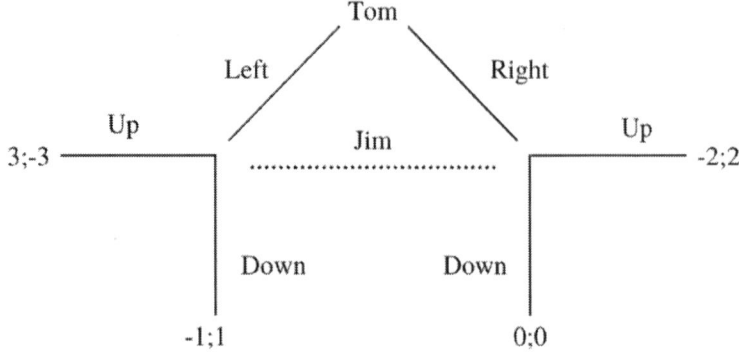

Picture 15[xli]: Scalped Subgame.

This is a full subgame so we can explore the Nash Equilibrium in this section and work our way up from this point. Because Jim lacks information about Tom's move, this becomes a *simultaneous* game, instead of a sequential one. If you read my previous book, Learn Game Theory[xlii], you already saw this game. It went by the name of a zero-sum mixed strategy game. We also encountered this game in Chapter 1, it is the Matching Pennies game with various payoffs. If you read Learn Game Theory, this chapter will refresh your memory on the mixed strategy algorithm. If you haven't read it, worry not. I will break down the game's solution and the mixed strategy algorithm.

In mixed-strategy games, players adopt a mixed strategy profile when they can't choose a definite action because there is no pure strategy equilibrium. They rather choose a course of action based on its *probability distribution*.

Say you're at a fast-food chain and can't decide whether to choose a salad or French fries with your meal. You are indifferent between the two options, but the cashier gets impatient by your slow decision making, so you flip a coin. Using this method, your probability of getting a salad is ½ or 50%. So is getting French fries, ½ or 50%. This was an easy probability calculation, but not all of them are so obvious.

Let's go back to our example above, where Jim is unsure whether Tom chose left or right. To make an informed decision in the absence of perfect information, Jim has to calculate the probability of Tom's choice using the payoffs. As the subgame is a simultaneous game, we can illustrate it in a game matrix:

Jim ↓ / Tom →	Left	Right
Up	3; -3	-2; 2
Down	-1; 1	0; 0

To solve this game, we will use an algorithm for mixed strategies to see which mixed strategies for each player make the other player indifferent.

First, we need to find a mixed strategy for Jim. He will play down in some cases, and up in others. We need to define a strategy for Jim where Tom will be indifferent to selecting left or right. If Tom gets the same payoff when he selects left or right, it won't matter to him which way to go. He can choose *at random* between left or right. And if Tom's random choice makes Jim indifferent between choosing up or down, it means Jim is satisfied with his original mixed strategy. Thus, neither of them will have any incentive to change their strategy. In other words, we will have a mixed-strategy Nash Equilibrium.

I know, this sounded like Albert Einstein talking in Dothraki. But bear with me. Everything will make much more sense as we delve into the solution of this problem.

The Mixed-Strategy Algorithm

Let's start with Jim's mixed strategy. We want to create a mixed strategy that makes Tom's *expected utility (*or payoff*)* for selecting left as a pure strategy equal to his expected utility (or payoff) if he goes right as a pure strategy.[xliii] We will mark expected utility as *EU*, left as $_L$, and right as $_R$. And we will illustrate everything we talked about in this paragraph as:

$$EU_L = EU_R$$

Remember that mixed strategies are based on probability. Each part of the equation is *a function* of a mixed strategy and represented in the following way:

$$EU_L = f(\sigma_U)$$
$$EU_R = f(\sigma_U)$$

In these two equations, f represents the function and σ_U (σ = sigma, $_U$ = up) shows the probability that Jim plays up. We have three equations with three unknowns!

Jim ↓ / Tom →	Left	Right
Up	3; -3	-2; 2
Down	-1; 1	0; 0

If Tom goes left, his payoff depends on Jim's decision—whether he will move up or down. If Jim goes up, Tom gets -3. If Jim goes down, Tom gets 1.

This is what $EU_L = f(\sigma_U)$ shows. The same stands true if Tom decides to go right. If Jim goes up, Tom gets 2. If Jim goes down, Tom gets 0. This is what this equation says: $EU_R = f(\sigma_U)$.

Jim ↓ / Tom →	Left
Up	3; **-3**
Down	-1; **1**

To solve for $EU_L = f(\sigma_U)$, we need to look at what Jim's expected utility for going left is as a function of the mixed strategy σ_U. Tom will get -3 for a percentage of the time and 1 for some percentage of the time (look at the bolded, underlined outcomes for Tom). The equation for the expected utility if Tom chooses to play left looks like this:

$$EU_L = \sigma_U(-3) + (1 - \sigma_U)(1)$$

To solve for Tom selecting left, sigma up is multiplied by -3 and added to 1 minus sigma up that has been multiplied by 1. Let's zoom in on each element of the equation above, one by one. Sigma up (σ_U) shows the probability of Jim playing up. The percentage of the time Jim plays up, Tom's outcome will be -3. We need to add this outcome to what happens the rest of the time. $1 - \sigma_U$ is the probability that Jim plays down. In this case, Tom is getting 1 as a payoff. So, we multiply the percentage of the time when Jim plays down (1-σ_U) with Tom's payoff: 1.

Jim ↓ / Tom →	Right
Up	-2; **2**
Down	0; **0**

We have to apply the same logic to the case when Tom chooses to move right. Let's write the equation for the expected utility of Tom going right:

$$EU_R = \sigma_U(2) + (1 - \sigma_U)(0)$$

What does this mean? This is the expected utility of going right as a function of sigma up (σ_U). Again, Jim goes up in some cases and then Tom's payoff is 2. On occasions, Jim will go down, and then Tom's payoff is 0. (See the bolded, underlined numbers of Tom.) As we see in the equation, sigma up (σ_U) shows the probability of Jim playing up. That percentage of the time, Tom's outcome will be 2. We need to add this outcome to what happens the rest of the time. 1 - σ_U is the probability that Jim plays down. In this case, Tom is getting 0 as a payoff.

So, we multiply the percentage of the time when Jim plays down $(1-\sigma_U)$ with Tom's payoff: 0.

We know from the first equation that $EU_L = EU_R$. Let's set both our equations up this way and work out the algebra:

$$\sigma_U(-3) + (1 - \sigma_U)(1) = \sigma_U(2) + (1 - \sigma_U)(0)$$

How do we solve this monster? I will guide you through it step by step. It is (not so) basic algebra.

Step 1: Simplify both sides of the equation.

$$\sigma_U(-3) + (1-\sigma_U)(1) = \sigma_U(2) + (1-\sigma_U)(0)$$

$\sigma_U(-3) + (1)(1) + (-\sigma_U)(1) = \sigma_U(2) + (1-\sigma_U)(0)$
(Distribute)

$-3\sigma_U + 1 + -1\sigma_U = 2\sigma_U + 0$

$[-3\sigma_U + (-1\sigma_U)] + (1) = (2\sigma_U) + (0)$ (Combine Like Terms)

$-4\sigma_U + 1 = 2\sigma_U$

$-4^{\sigma_U} + 1 = 2^{\sigma_U}$

Step 2: Subtract 2^{σ_U} from both sides.

$-4^{\sigma_U} + 1 - \mathbf{2^{\sigma_U}} = 2^{\sigma_U} - \mathbf{2^{\sigma_U}}$

$-6^{\sigma_U} + 1 = 0$

Step 3: Subtract 1 from both sides.

$-6^{\sigma_U} + 1 - \mathbf{1} = 0 - \mathbf{1}$

$-6^{\sigma_U} = -1$

Step 4: Divide both sides by -6.

$-6^{\sigma_U} / \mathbf{-6} = -1 / \mathbf{-6}$

$\sigma_U = 1/6$

Answer:

$\sigma_U = 1/6$

  This tells us that when Jim plays up 1/6 of the time and down 5/6 of the time, it doesn't matter if

Tom plays left or right. Regardless, Tom will have the same expected utility.

We can follow the same steps as above to find a strategy that will make Jim indifferent to Tom's decision to move left or right. We ask, what's Jim's expected utility for up or down when those two things are equal? Let's illustrate it like this:

$$EU_U = EU_D$$

The expected utility of playing up is a function of Tom's choice of playing left.
$$EU_U = f(\sigma_L)$$

Similarly, the expected utility of playing down is a function of Tom's decision of playing left.

$$EU_D = f(\sigma_L)$$

Let's bring everything together:

Jim ↓ / Tom →	Left	Right
Up	3; -3	-2; 2

Jim's expected utility to play up is 3 if Tom moves left and -2 if Tom moves right. See the equation below:

$$EU_U = \sigma_L(3) + (1 - \sigma_L)(-2)$$

Jim ↓ / Tom →	Left	Right
Down	-1; 1	0; 0

Jim's expected utility to play down is -1 if Tom moves left and 0 if Tom moves right.

$$EU_D = \sigma_L(-1) + (1 - \sigma_L)(0)$$

As $EU_U = EU_D$, we can write our equation as follows:

$$\sigma_L(3) + (1 - \sigma_L)(-2) = \sigma_L(-1) + (1 - \sigma_L)(0)$$

Let's solve this equation step-by-step.

Step 1: Simplify both sides of the equation.

$$\sigma_L(3) + (1 - \sigma_L)(-2) = \sigma_L(-1) + (1 - \sigma_L)(0)$$

$$\sigma_L(3) + (1)(-2) + (-\sigma_L)(-2) = \sigma_L(-1) + (1 - \sigma_L)(0)$$
(Distribute)

$$3\sigma_L + -2 + 2\sigma_L = -\sigma_L + 0$$

$(3\sigma_L + 2\sigma_L) + (-2) = (-\sigma_L) + (0)$ (Combine Like Terms)

$[5\sigma_L + (-2)] = -\sigma_L$

$5\sigma_L - 2 = -\sigma_L$

Step 2: Add σ_L to both sides.

$5\sigma_L + -2 + \sigma_L = -\sigma_L + \sigma_L$

$6\sigma_L - 2 = 0$

Step 3: Add 2 to both sides.

$6\sigma_L - 2 + 2 = 0 + 2$

$6\sigma_L = 2$

Step 4: Divide both sides by 6.

$6\sigma_L / 6 = 2 / 6$

$\sigma_L = 1/3$

Answer:

$\sigma_L = 1/3$

The result is that when Tom has the same expected utility and plays left with the probability 1/3 of time and right with a probability 2/3 of the time, Jim is indifferent when he plays up or down.

Let's fill our game matrix with all the information we got. So, in this game, the mixed-strategy Nash Equilibrium is when Jim plays up 1/6 of the time, and Tom plays left 1/2 of the time.

$\sigma_U = 1/6$ and $\sigma_L = 1/3$

Jim ↓ / Tom →	Left 1/3	Right 2/3
Up 1/6	3; -3	-2; 2
Down 5/6	-1; 1	0; 0

As long as Tom and Jim stick to these strategies, neither can change what they do and expect better results. Why? Because Jim's expected utility for playing up and down are the same, and so is Tom's for playing left or right.[xliv] In other words, the unique Nash Equilibrium of this game is when Jim goes up with a probability of 1/6 and down with a probability of 5/6, Tom goes left with a probability of 1/3 and right with a probability of 2/3.

But let's not forget that this calculation doesn't solve the entire game. In Picture 15, we removed Jim's first step. Namely, whether he should go or stay. So now we should explore Jim's expected utility for going. Thanks to the mixed strategy algorithm, we know that Tom will mix between going left and right, and Jim will mix between going up and down. Now, we need to calculate the quality of the "Go" outcome for Jim and compare it with how much he will get if he "Stays."

How do we calculate payoffs?

So far we calculated what Jim and Tom do in a mixed-strategy Nash Equilibrium. However, we don't know the actual payoffs of their mixed strategies. Now we will learn how to calculate it!

What we know:
- if Jim goes Up and Tom goes Left, Jim gets 3 and Tom gets -3.
- If Jim goes Down and Tom goes Left, Jim gets -1 and Tom gets 1.
- If Jim goes Up and Tom goes Right, Jim gets -2 and Tom gets 2.
- If Jim goes Down and Tom goes Right, Jim gets 0 and Tom gets 0.

What we don't know:
- What the exact number for the mixed-strategy Nash Equilibrium is. All we know are the probabilities:

Jim ↓ / Tom →	Left 1/3	Right 2/3
Up 1/6	3; -3	-2; 2
Down 5/6	-1; 1	0; 0

To make our calculation, we will need to use another algorithm.

1. First, we need to find the probability that each outcome happens in equilibrium.
2. Then, we need to multiply that probability with a given player's payoff for each outcome.
3. Sum all these numbers together.[xlv]

This calculation won't be so complicated as it sounds, I promise. How do we eat an elephant? By chopping it into bite-sized pieces. Let's start with bite number one.

1. First, we need to find the probability that each outcome happens in equilibrium.

Jim ↓ / Tom	Left <u>1/3</u>	Right <u>2/3</u>

	→	
Up **1/6**	3; -3	-2; 2
Down **5/6**	-1; 1	0; 0

We've already found the probabilities of the players. See the bolded and underlined numbers. The next step is to multiply across the matrix. To see the probability of Jim going Up and Tom going Left, you take Jim's probability, 1/6, and multiply it with Tom's probability, 1/3.
1/6 x 1/3 = 1/18

Now, we do the same calculation for each scenario.

Jim Down - Tom Left = 5/6 x 1/3 = 5/18
Jim Up – Tom Right = 1/6 x 2/3 = 2/18
Jim Down – Tom Right = 5/6 x 2/3 = 10/18

Let's add our results to the matrix:

Jim↓/To→	**Left 1/3**	**Right 2/3**
Up 1/6	3; -3 → 1/18	-2; 2 → 2/18
Down 5/6	-1; 1 → 5/18	0; 0 → 10/18

We could say that the payoff for Jim going Up with a probability of 1/6 and Tom going Left with a probability of 1/3 is 1/18, and so on.

The next step is to take the probabilities and multiply them with their outcomes. Let's start with Jim's payoffs: [xlvi]

Jim ↓ / Tom →	Left 1/3	Right 2/3
Up 1/6	3 x 1/18	-2 x 2/18
Down 5/6	-1 x 5/18	0 x 10/18

After multiplying each of Jim's values in the matrix, we get this calculation:

3/18+(-4/18)+(-5/18)+0 = -6/18 = **-1/3**

Now let's calculate Tom's payoffs in the subgame:

Jim ↓ / Tom →	Left 1/3	Right 2/3
Up 1/6	-3 x 1/18	2 x 2/18
Down 5/6	1 x 5/18	0 x 10/18

Tom's values in the matrix will give us this calculation:

-3/18+4/18+5/18+0=6/18=**1/3**

Where do these numbers belong? These are the final payoffs of the subgame illustrated in Picture 15. These are the expected payoffs if Jim decides to "Go."

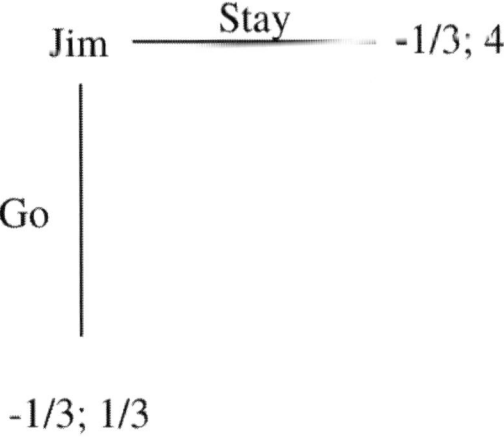

Picture 16[xlvii]: The top of the game tree – Jim's outcome if he decides to go.

What we can see in Picture 16 is that Jim's payoff is -1/3 regardless of his choice of going or staying. Tom's payoffs would be 1/3 if Jim goes and 4 if Jim stays, but Tom doesn't have a move opportunity yet. The only information that matters here is that Jim is indifferent between staying and going as he'll get the same payoff, -1/3. In other words, there are *multiple subgame perfect equilibria* in this game.

What does this mean?

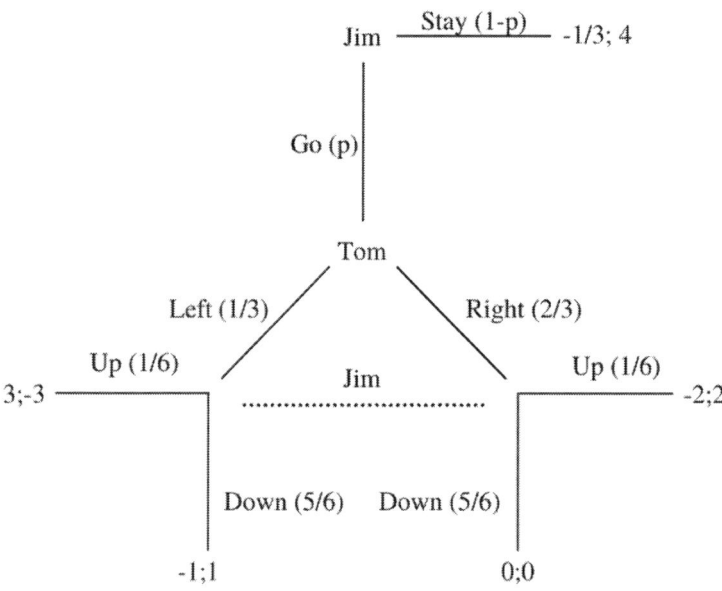

Picture 17: Multiple Subgame Equilibria[xlviii]

What we see on this game tree is that Jim can choose "Go" with probability (p) and "Stay" with probability (1-p). (P) can be any number. If (p) is 1, Jim always chooses "Go" and plays it as a pure strategy. If (p) is 2, then Jim "goes" twice as often as he "stays," and so on. There are as many options for Jim as numbers – infinitely many.[xlix] Then, when he "goes," he plays Up with a probability of 1/6 and down with a probability of 5/6. Tom plays Left with a probability of 1/3 and Right with a probability of 2/3.

When you need to find the Nash equilibria of a game, you first look for pure strategies. If you can't

83

find them, that's when you explore mixed strategies. To find the probabilities of the expected payoffs, you have to run the mixed strategy algorithm. Then you need to calculate those payoffs. Then you check if the Nash Equilibrium is also subgame perfect (meaning it stands true with all subgames and the main game). The sign that you're dealing with multiple subgame perfect equilibria is the existence of an indifference – in our case, Jim's Go-Stay choices both giving exactly -1/3 as a payoff.

Exercise 1: The Battle of the Sexes calculation.

As I promised, the Battle of the Sexes is yours to calculate. This exercise is somewhat aided as you know what percentages you need to get after running the algorithm and calculating the payoffs. As a reminder, here is the matrix:

Jim ↓ / Kat →	Dance	Rugby
Dance	1; 2	0; 0
Rugby	0; 0	2; 1

a.) Run the mixed-strategy algorithm.

b.) Calculate the payoffs.

Exercise 2: Health care problems.

This exercise will be harder as the results are not displayed. Also, this game is tricky. The results will surprise you. Just for practice's sake, follow the calculations just as you learned in this chapter until the very end of the payoff calculation. In the end, it will make sense why is the mutually suboptimal choice the only choice Jim can make. You can find the results and explanations at the end of this book.

Jim ↓ / ER MD →	Quick	Thorough
ER	3; 5	5; 4
Specialist	4; 0	4; 0

a.) Run the mixed-strategy algorithm.
b.) Calculate the payoffs.

Chapter 5: Game Stacking

How often did it happen to you that you played only one round of rock-paper-scissors? Few people stop at just one game. Usually, you play at least three rounds, and whoever wins only one will be in charge of washing the dishes after dinner. Sometimes you and your friends play multiple rounds of rock-paper-scissors just because it's fun.

So far, we assumed that our players, Jim and Tom, play only one round of, say, the Prisoner's Dilemma, or the Stag Hunt, or the Battle of the Sexes. But what if they played multiple rounds of each of these games? Or, what would happen if they played them one after the other? How would the chain of these games flow? Would the first game they play influence the second?

Even in a simple game such as rock-paper-scissors, you strategize. If your partner played rock before and lost, you may think this time they will play something else and try winning that way. Or, quite the opposite, what if they play rock again thinking that you will play something other than paper thinking that they will play something other than rock?

Let's say Jim and Tom play three games. They start with a Prisoner's Dilemma, collect their payoffs, then play the Battle of the Sexes, collect the payoffs, and finally, they have a Stag Hunt match. The winner of this three-stage game will be the one who collects the large collective payoff.

There can be finitely or infinitely many stages and variations of these games. In game theory, we call them finitely or infinitely repeated games. For the scope of this book, we will stick with finite games. Players can play six rounds of just the Prisoner's Dilemma, or just two rounds, one being Stag Hunt the other being the Matching Pennies, for instance. Each individual game is a subgame within the overall, larger game.

While one can mix and match the subgames and the number of rounds, some attributives apply to each game played:

- The games are simultaneous.
- The payoff of one game doesn't influence the payoff of another game.
- Each player knows the moves of other players.[1]

The players know each other's moves not only in the particular subgame of the moment, but also retrospectively. Say, Jim and Tom play six Prisoner's Dilemmas. In the fifth subgame, both will have information about the previous four moves the other player chose. As such, players will aspire to play in subgame perfect equilibrium.

These stacked games would be illustrated on a game tree, but it would take up a lot of space and time to do it. Why? For example, if we played only two rounds of the Prisoner's Dilemma, we must draw out the four outcomes the game has, and then we would need to add another Prisoner's Dilemma to each outcome. This would give sixteen outcomes in the second round. If we'd add a third round, we'd have to attach four outcomes to the already existing six. So, imagine how it would multiply if we tried to illustrate all the six rounds of the game… Clearly, a game tree is not a realistic solution here. So then what is? William Spaniel, a game theory expert, talks about two theorems we can summon to help us with analyzing stacked games.[li]

The first theorem states that in the last subgame, all players must play a Nash Equilibrium in all subgame perfect equilibria. This means that in our last Prisoner's Dilemma game, the players must play the Betray-Betray strategy. If the final subgame is a

Stag Hunt, the players must play either the Hare-Hare, Stag-Stag strategies, or the unique mixed strategy of the game.

You may ask, "why is playing a Nash Equilibrium so important in the last stage?" Spaniel explains that whatever payoffs you earned in all previous subgames of the main games are unchangeable, and they are in the past. Players can only influence their payoffs in the present subgame, so they better try to maximize it. And a Nash Equilibrium is a state when no player has the incentive to deviate from their strategy. This means there is no other strategy that would give a player a higher payoff. In other words, a Nash Equilibrium yields the biggest payoff for every player.

The second theorem states that playing a Nash Equilibrium in each subgame of the main game is a subgame perfect equilibrium. What does this mean? If players played two rounds, one Prisoner's Dilemma and one Stag Hunt, they'd play the Nash Equilibrium in both games, not just the last one. So, in the Prisoner's Dilemma, they'd play the only Nash Equilibrium, the Betray-Betray option. And in the Stag Hunt, they'd choose, say, the Stag-Stag option. Each player maximized their payoffs in both games, meaning they played Nash Equilibria, and thus the game is also a subgame perfect equilibrium. The same would stand true if the players chose the Hare-Hare or the mixed-strategy option in the Stag Hunt game. So, this two-stage game has three subgame perfect equilibria as a solution: Betray-Betray + Stag-Stag, Betray-Betray+Hare-Hare, and Betray-Betray+Mixed-Strategy Nash Equilibrium.

If we played two Stag Hunts, we'd have three Nash Equilibria at the end of the first game, and three Nash Equilibria linked to each of the first three after playing the second game, so a total of nine subgame perfect equilibria outcomes. So, as you can see, there can be a lot of subgame perfect equilibria after multiple round games.

Our previous payoffs won't affect future payoffs when we play like this. We try to maximize each stage in the moment we play them. Thus, we must choose the Nash Equilibria.

If, however, we scrutinize the definition of theorem 2, we notice it says, "playing a Nash Equilibrium in each subgame of the main game is *a* subgame perfect equilibrium" - but not the only one. If players adopt strategies that consider and respond to previous plays, more equilibria are possible.

Playing Nash Equilibria in each round seems smart in theory, but would it hold true in practice. If someone wronged you yesterday, would that wrong be erased today? Would you play a game without yesterday's harm in mind? Is it realistic to say that each game round is separate then, one not affecting the decision-making of the other? Well, we will see that in real life, the answer is no.

In real life, players form strange alliances over time and cooperate where it doesn't make mathematical sense based on what we have learned so far. For example, in a one-round Prisoner's Dilemma, players would rat out one another. But maybe in the second round, after spending 8 years in prison, they would more likely cooperate and stay silent, even if individual deviation would bring a better outcome for each. Or, players may do the opposite and take revenge even when it's not in their best interest. [lii]

The Purpose of Punishment[liii]

Let's assume that we're playing a two-stage game, a Prisoner's Dilemma first and then a game called Free Money Game. Before I illustrate how the sequence of these two games can play out, let me briefly present the Free Money game, a unique, *even number* Nash Equilibria game.

In this game, Jim and Tom are offered money. The only thing they have to do is unanimously, simultaneously, and blindly approve receiving the free money. If Jim and Tom both say yes to getting the money, they both will. But if one of them says no, neither gets any money.

I know, who on earth would say no to free money? Remember this question for later – it will be important when we discuss punishment strategies! But for now, let's analyze this game.

Jim ↓ / Tom →	Yes	No
Yes	5; 5	0; 0
No	0; 0	0; 0

This is what the free money game looks like. As you can see, if both Jim and Tom say yes, they get five dollars. If either or both say no, they get nothing. Yes-yes, therefore, is a pure-strategy Nash Equilibrium. The brothers can't change their strategy and get a better outcome.

Jim ↓ / Tom →	No
No	0; 0

But no-no is also a pure-strategy Nash Equilibrium because they can't profitably deviate from their response. If only Jim said yes, they would still get nothing. The same is true if Tom says yes.[liv]

So how would the Prisoner's Dilemma and the Free Money game interact if played as a sequence?

Game 1: The Prisoner's Dilemma

Jim ↓ / Tom →	Stays Silent	Betrays
Stays Silent	4 ; 4	1; 5
Betrays	5; 1	3; 3

I added new numerical values to the Prisoner's Dilemma game. As you can see, the only pure strategy Nash Equilibrium is still the Betray-Betray option. Why? Because if both would stay silent, individually, each could have a profitable deviation. If Jim and Tom agreed that no matter what, they would stay silent, but last-minute Jim changed his mind and betrayed Tom, as he would get a higher individual payoff, 5 instead of 4. Tom would lose some of the payoffs, he would get 1 instead of 4. The same is true if Tom decided to betray.

And betray they would if this was a one-stage game. There would be no consequences. There is, however, a second game coming.

Game 2: The Free Money Game.

Jim ↓ / Tom →	Yes	No
Yes	5; 5	0; 0
No	0; 0	0; 0

Let's assume Jim betrayed Tom despite their agreement, so now Tom is 3 points short. How likely are you to trust and cooperate with the Betrayer the next round if this ever happened to you? Probably unlikely. Tom might say "No" in the Free Money game just to punish Jim. And Jim knows that. The second game acts as a deterrent for playing nasty in the first game. Knowing that Tom might punish him in the next game, Jim is less likely to deviate from cooperation. They both want to maximize their payoffs, so they will play the Nash Equilibrium with the highest reward, which, in this case, is staying silent. Cooperating successfully in the first round builds trust, so they will cooperate in the second round, as well, both maximizing their payoffs to a total of 4+5=9. If Jim didn't cooperate, he would have gotten a payoff of 5 and Tom a payoff of 1.

So back to your question, "who on earth would say no to free money?" Well, someone who just got played in a previous game is now looking for revenge. Or better yet, the sheer possibility of a punishment present deters players from unilaterally deviating from a cooperative strategy.

In game theory terms, in the first game, the Prisoner's Dilemma, players play the strictly *dominated* strategy in cooperation, because the next game promises a reward if they do so. Thus, we have a subgame perfect equilibrium that doesn't follow the obvious, pure strategy Nash Equilibrium. The punishment strategy acts as a threat, a deterrent, and keeps players in check. Think about the Cold War. Both the US and the USSR had nukes. Neither country wanted to actually use them, but these weapons of mass destruction acted as a buffer for each country to play -relatively- nicely. This is not a perfect analogy but works on the same principle of deterrence.

The bottom line is that a lot of strategic decision-making can go into repeated games. Think about examples from your real life. Do you tend to go to the same hairdresser, car mechanic, or even farmer at the farmer's market? Why? Because over time, due to repetition, you built some trust with them, and you know that they do a good job. In this "strategic environment," you know that choosing them will be beneficial for you. To build beneficial long-term relationships requires trust-building over time, and repeated games are a way to analyze this process in mathematical terms.[lv]

What we intend to capture with repeated games is that after ten, twenty, or fifty rounds can we trust the other player to make a move that will be mutually beneficial. Repeated games are perfect for analyzing the interaction between immediate payoffs and long-term incentives. For example, if your hairdresser thinks about overcharging you for the haircut, they may gain more from that one interaction, but they may lose you as a customer. Because of the threat of future losses, they are incentivized to give you a great financial deal and service, so you come back to their salon. When the long-term benefits outweigh immediate gains, the strategic environment changes. This is not something one thinks about in stand-alone games.

However, beyond committing to something, it is essential to *credibly* commit to build trust with another player. In the next chapter, we will explore the difference between credible and non-credible threats and commitments.

Chapter 6: Credibility and Non-Credibility

My grandfather lived in a small village in Poland by a riverbank in the 1940s. He fought in World War II. He had many war stories, but the following was my favorite. When the tides turned, the Soviets started pushing back the German troops. The Germans invaded our village, but upon hearing that the Soviets were coming, the invaders gathered their forces to relocate to a strategically more sensible place to meet the Russian troops.

Our village was located next to a fast-flowing, wide river with a double bridge and a little island in the middle. If I drew out a primitive map, this is how it looked:

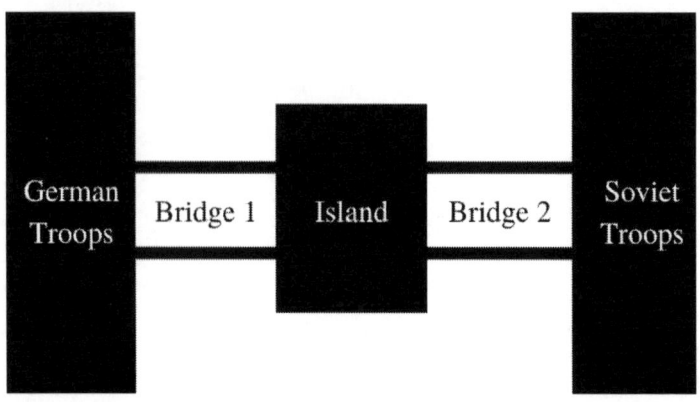

Picture 18: Map of military positions.

The Germans stayed in our village, to have access to food supplies and took control over Bridge 1. The Soviet troops were on the other side of the river, having access to Bridge 2. The island was a type of no man's land, not a valuable asset to possess. So what happens next?

First, we will examine the Game Theory solution to this game, and then I will tell you how the story actually ended.

If players face this set up, even though the island has no resources, it's still better to have it than not to have it and allowing the enemy to occupy it. Say the Soviets move first. They relocate their troops to the island and can choose between two options, burning the bridge behind them, or not burning the bridge.

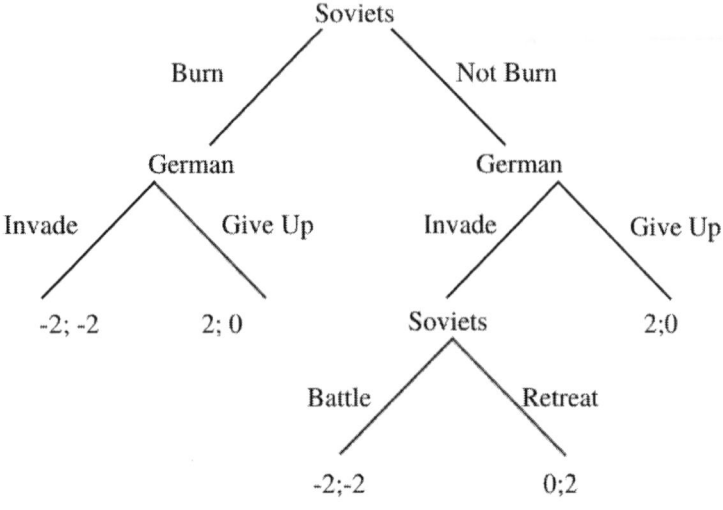

Picture 19[lvi]: War Zone.

When a player wants to deliver a credible threat, they must be ready to do it by making a point. How? Well, Player 1, in our case, the Soviets, decided to make their first move. They marched their troops over Bridge 2, spreading out their army on the small island. Now, they have two choices: they can either burn the bridge behind them, leaving no other option for themselves but to advance over to the German territory, or they can choose not to burn the bridge, keeping an escape route if things don't go their way.

On the game tree in Picture 19, we can see that if the Soviets choose not to burn the bridge, the Germans can either give up or invade, moving their troops towards the island. If Germans give up without a fight, their payoff will be 0 – they lost no soldiers (let's just assume they wouldn't end up on the Gulag as war hostages). The Soviets would get 2 as a payoff, they wouldn't lose any soldiers either and would dictate the next steps of the war.

With a German invasion, Soviets would have two options. They could either engage in battle that would leave both troops with severe losses, and a payoff of -2 and -2. Or the Soviets could retreat and with no other losses than pride and moral, they would get a 0. The Germans, gaining control over the island without a battle, would receive a payoff of 2.

If, however, the Soviets go the other route and burn the bridge, the Germans would have no other option than to engage the Soviets one way or another. They could choose to invade the Soviets now trapped on the island, both getting a payoff of -2. Or they could give up and gain 0 as payoff, while the Soviets would get 2. How to solve this game? Let's recall what we learned about backward induction – we go to the last move of a give game and work our way back from there.

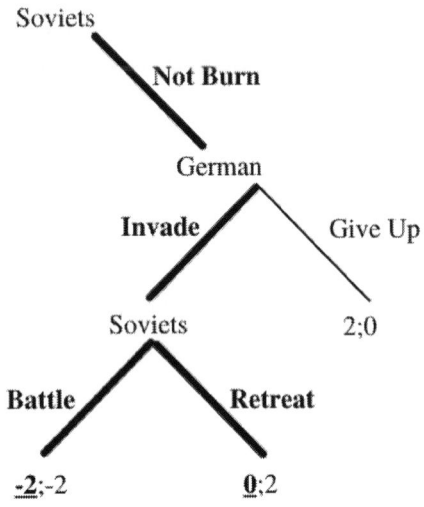

Picture 20: The right side of the game tree.

If the Soviets don't burn the bridge and the Germans invade, we can see that the Soviets will have these two responses: engage in battle for a payoff of -2 or retreat for a payoff of 0. Knowing that our players in game theory are rational decision-makers, a 0 is a better outcome than a -2. So, the Soviets would choose retreat instead of a battle if the Germans called the Soviets' bluff and invaded. It would be like, never mind, just kidding, we will go back to our shore.

Will the Germans always invade? Yes. Why? Because, if you compare the German outcome for giving up (0) and the German outcome for invading and the Soviets retreating (2) in Picture 20, it's clear for the Germans they will be better off invading given the Soviets didn't burn the bridge. The Germans will get a higher payoff (2 vs. 0) by doing so. It makes no sense for the Germans to give up when the Soviets don't burn the bridge.

Let's analyze the second scenario now where the Soviets burn the bridge.

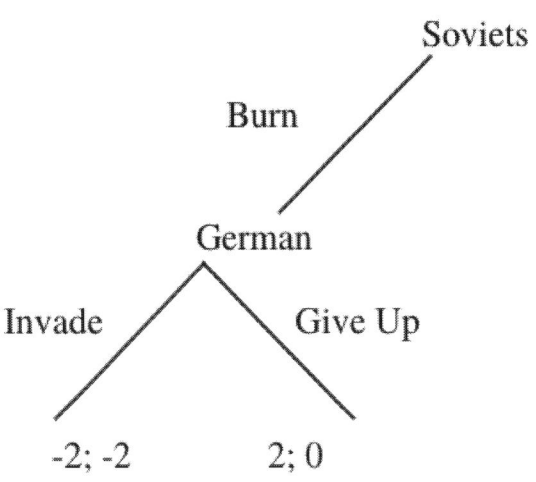

Picture 21: The left side of the game tree.

When the Soviets do burn the bridge, the Germans can either invade with a payoff of -2 or retreat with a payoff of 0. Understanding that 0 is better than -2, in case of a burned bridge, the Germans will always choose giving up.

This is not quite the end of our analysis. We need to compare the Soviets' two final outcomes to conclude whether they should or shouldn't burn the bridge.

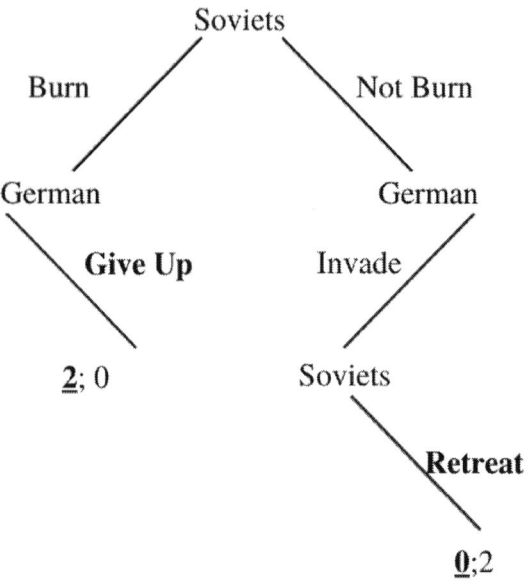

Picture 22: Outcome analysis.

Looking at Picture 22, it's clear that the Soviets are better off burning the bridge. If they do so, the Germans will be forced to give up, so the Soviets will gain 2. However, if the Soviets don't burn the bridge, the Germans will invade, forcing them to retreat with a payoff of 0. 2 is better than 0, so the Soviets will choose to burn the bridge behind them. The very act of burning the bridge and forcing a limit to future actions makes the Soviets' threat credible.

This game is meant to illustrate that limiting future actions can help make a threat credible and corner players into choices they may not make otherwise.[lvii]

I used my grandfather's story of the Germans and the Soviets as the situation fits well with this game theory example. However, reality played out differently. The Germans in our village chose to blow up the bridge on their side, so Bridge 1 in Picture 18. They did this because their troops were already decimated, and knew they wouldn't stand a chance if they had to have a face to face combat with the Soviets. So, they blew up the bridge on their side to win some time and join forces with some other German troops in the area. They took what they could from our village, burnt our fields, and ran away. The Soviets took a couple of weeks to build another bridge to cross the river. When they crossed, they did what invaders do, took whatever little the Germans left, and continued their hunt.

Non-Credibility

Let's refresh our memory on the Health Care Problems game. Remember, this was the game where Jim had a medical problem, and he could choose between going to the ER immediately vs. waiting to see a specialist. Due to the conflicting interests of Jim and the ER doctor, it turned out that waiting for the specialist examination will always be better for Jim than going to the ER as the ER doctor can't credibly commit to performing a thorough health examination once Jim is in the emergency room.

The Health Care Problems game could be rewritten with the following plot: why would locals always revolt against invaders? Let's switch the name of the players from Jim and ER MD to Locals and Invaders and go back to our village in Poland while the German troops still occupied them.

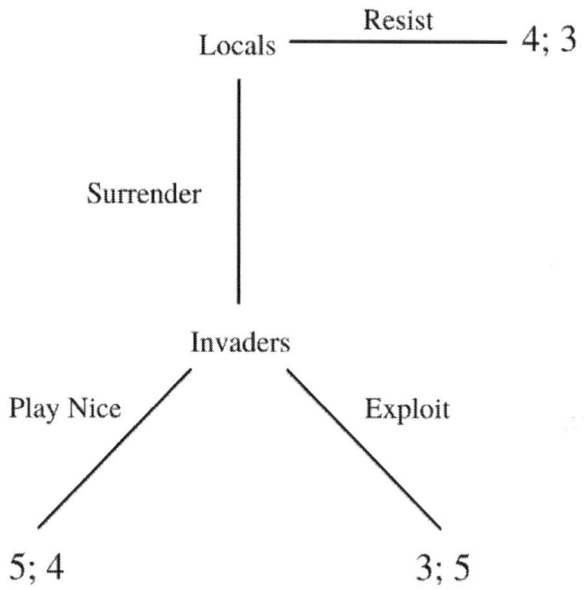

Picture 23: Locals and invaders.

At the beginning of World War II, the Germans occupied Poland swiftly and aggressively. The locals had new rules and limitations imposed on by the invaders, which made them resentful and non-cooperative. If we look a Picture 23, we can see that the locals can give two responses to the invader's new rules – surrender and cooperate with the invaders or fight the new rules in any way they can. If they resist, they gain advantages, they don't get totally exploited (payoff 4). The invaders gain resources, but it costs them extra effort to keep the locals in check (payoff 3).

If the locals surrender, the invaders could play nice with the locals, establish a somewhat friendly relationship and gain a payoff of 4. The locals would be best off as the invaders wouldn't harm them, getting a payoff of 5. However, invaders being invaders with big guns, tanks, and whatnot, will likely not play nice but exploit the locals instead, gaining the highest payoff 5, while the locals will be under severe distress, gaining a payoff of 3. So, it is always better for the locals to choose the resist option instead of a complete surrender, even if they could have a better outcome if they surrendered and somehow convinced the invaders to play nicely. The invaders can't credibly commit to that option. This story follows the same logic as the Health Care Problems game.

If, somehow, the locals would overpower the invaders, the roles would change, and Invaders would become Player 1 and locals Player 2, the dynamic would persist. The locals wouldn't be able to credibly commit to not exploiting the invaders in case they had the upper hand, so the invaders would also resist and fight instead of surrendering and hoping that the locals would give them what they needed, so they could be on their way.

Conflicts – and games - of such nature can go on in perpetuity. Power dynamics can shift, the players can switch places, but the dynamic will persist. This is a symmetrical game until one player is annihilated.

Many steps, one conclusion

The Centipede Game[lviii]

The Centipede Game is another famous game in game theory. This game is also a good illustration of the problems arising from non-credible threats and the inability of credible commitments. The game goes as follows:

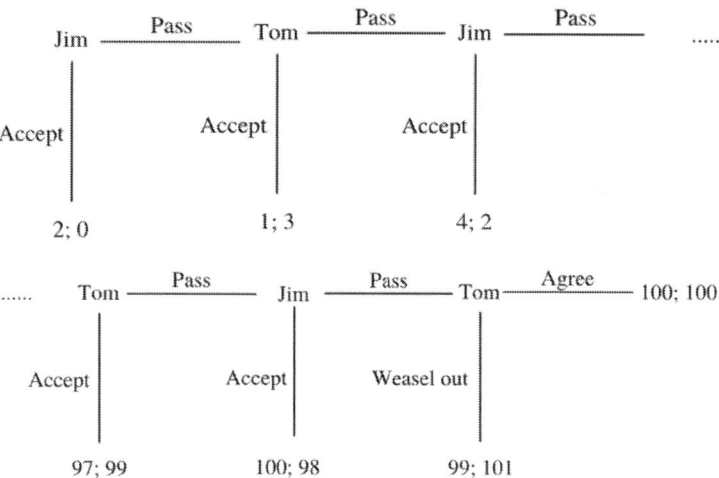

Picture 24: The Centipede Game

In this game, Jim and Tom are promised candies. They take turns sequentially, and both can either accept the offer or Pass. Whenever they Pass, the other player's potential candy gains will increase by 2 in the next round. Whenever they accept, the accepting player gets two candies more from the jar - the other player gets 2 less. In the end, the last player can either weasel out of the game or be kind and split the candies in half. If Jim accepts the candy offer in the first round, he'll get 2, and Tom will get 0. If Jim Passes on the first round, Tom could accept the candy offer, now Tom getting 3 candies and Jim getting 1. Tom could also Pass on the second offer, throwing the proverbial ball into Jim's court to accept or Pass. Fast forward 98 steps and we are in the last round. Here Tom could weasel out of the final candy offer, gaining 101 candies leaving Jim with 99. Or he could agree to honor the game's last move, both gaining 100 candies.

Wow, you may think. 100 steps? It will take forever to solve this game. Even with backward induction! It's still 100 steps! Worry not; this game is not as scary as it looks.

How to solve this game? Using backward induction and assuming sequential rationality, we go to the last move where Player 2, Tom, has to decide between weaseling out for a payoff of 101, or agreeing for a payoff of 100. As 101 is greater than 100, Tom will choose to weasel out. We take a step back. It is Jim's turn. He will have two options: either accept the candy offer and gain 100, leaving Tom with 98, or Pass and gain 99 – Jim knows that Tom will choose Weasel Out in that round. 101 is bigger than 99, so Jim, being a rational decision maker will Accept instead of Pass. Tom's turn again. He can choose between accepting and gaining 99, or Passing and getting 98 as he knows Jim will choose Accept in the next round.

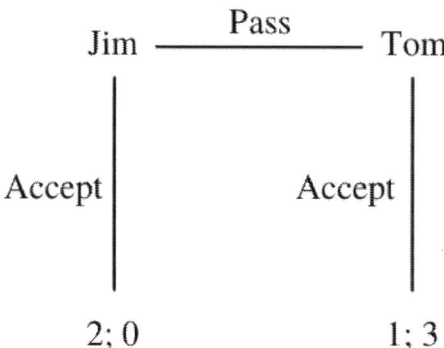

Picture 25: The first/final step.

Rewind to the last (or better first) step of the game. It's Jim's choice to Accept the 2 candies, leaving Tom with nothing, or Pass and gaining 1 candy, knowing that Tom will accept the next round. Jim will compare 2 to 1 and choose Accept instead of Passing.

It doesn't matter how many steps there are in the game, each move follows the same strategy, and you can quickly catch up on the dynamic. Instead of analyzing each step individually, you can fast forward to the first step and see what makes sense there.

How to find the subgame Perfect Equilibrium?

In the Introduction, I promised that we would learn how to identify a game's subgame perfect equilibrium in a sequential multi-step game. To make things easier, I will transform the Centipede Game into a Quadruped Game.

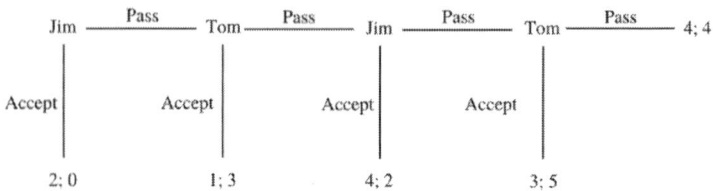

Picture 26: The Quadruped Game

One Nash Equilibrium would be for each player to choose Accept in each of the four stages, or (AA; AA) where the first and second A are Jim's choices for round 1 and 3. The third and fourth A represent Tom's choices in round 2 and 4. But this is not the only Nash Equilibrium. In fact, there are four

Nash Equilibria. As long as Jim and Tom both choose "Accept" at least in one of their moves, it is a Nash Equilibrium.[lix]

The 4 Nash Equilibria:
(AA; AA) – Both Jim and Tom choose Accept each time.
(AA; AP) – Jim chooses Accept in round 1 and 3. Tom chooses Accept in round 2 and Pass in round 4.
(AP; AA) - Jim chooses Accept in round 1 and Pass in round 3, Tom chooses accept in both round 2 and 4.
(AP; AP) – Jim chooses Accept in round 1 and Pass in round 3. Tom chooses Accept in round 2 and Pass in round 4.

Another way to illustrate this is to put both Jim's and Tom's strategies into one equation:
$S_J = S_T = \{AA, AP, PA, PP\}$

S_J stands for Jim's strategy, and S_T stands for Tom's strategy. The equation means that both players have the same strategy profile, they can both Accept, both Pass, and Accept and Pass, or Pass and accept. So, depending how far in the game they got, they both could think that they will Accept their first round and Pass the second. Obviously, if they accept their first round, their second round will not happen, but we can hypothetically analyze possible steps within a set of information.

We can illustrate this game in a matrix:

4;4	3;5	1;3	1;3
4;2	4;2	1;3	1;3
2;0	2;0	2;0	2;0
2;0	2;0	2;0	2;0

In this table, we see Jim's and Tom's choices in a game matrix – or normal form. In the first row, we see that, if Tom knows that Jim will Pass both rounds 1 and 3, what's the best course of action for Tom to take? There, it's better for him to Pass round 2 and then Accept round 4. This way, Tom will get a payoff of 5 instead of 4. So, the best response for Tom is to play Pass first and Accept second if Jim chooses the Pass-Pass strategy.

After analyzing each combination for AA, AP, PA, and PP, I bolded and underlined the optimal responses of both Jim and Tom in the table below.

4;4	3; 5	1;3	1;3
4;2	**4**;2	1;**3**	1;**3**
2;**0**	2;**0**	**2;0**	**2;0**

2;0	**2**;0	**2;0**	**2;0**

The four Nash Equilibria, therefore, where neither player can deviate from their current choices are: {(AA, AA); (AP, AP); (AA, AP); (AP, AA)}

Now that we got our results for the full game, let's remove Jim's first round, and look at the three-round subgame:

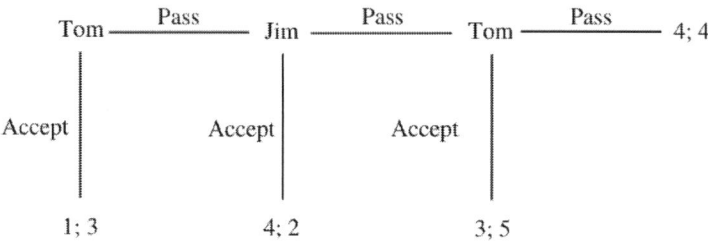

Picture 27: The first subgame of the Quadruped Game.

In this subgame, Jim has only one round, which we can illustrate as such:

S_J= {-P; -A}, where S_J stands for Jim's strategy and -P and -A show he has one less move.

S_T= {PP, PA, AP, AA}

4;4	3; 5	1;3	1;3

| 4;2 | 4;2 | 1;3 | 1;3 |

If we illustrated the variety of the moves in normal form, this is what we would get. If we write down the best responses of each player in this subgame, we get two Nash Equilibria:

| 4;4 | 3; 5 | 1;3 | 1;3 |
| 4;2 | 4;2 | **<u>1;3</u>** | **<u>1;3</u>** |

The two underlined and bolded outcomes are when Tom plays Accept and Accept (AA) when Jim plays Accept (-A) in his second round, and Tom plays Accept and Pass when Jim plays Accept (-A) in his second round. The Nash Equilibria are: {(-A, AA); (-A, AP)}.

What we can notice here is, that if we compare the Nash equilibria of the subgame with the Nash Equilibria of the original game, only two are a consistent match[lx]:

The Nash Equilibria of the subgame: {(-A, AA); (-A, AP)}.
The Nash Equilibria of the original game: {**(AA, AA)**; ~~(AP, AP);~~ **(AA, AP)**; ~~(AP, AA)~~}

To have subgame perfect Nash Equilibria, we learned that the whole game's subgame perfect equilibrium has to stand true *in all the* subgames as well. So, we will eliminate the main game's Nash Equilibria that don't stand true in the subgame.

Now, let's analyze the second subgame:

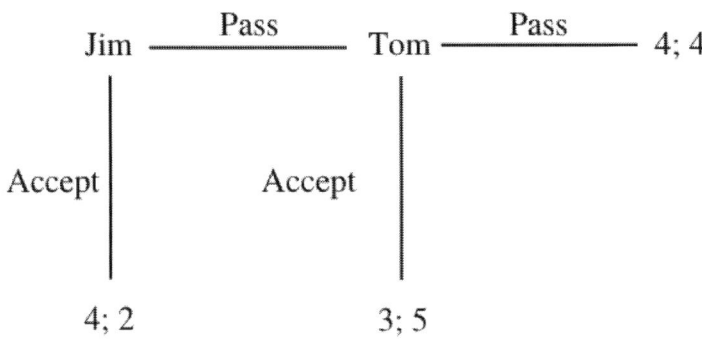

Picture 28: The second subgame.

In this subgame, Jim and Tom both have only one round, which we can illustrate as such:

$S_J = S_T = \{-P; -A\}$, where S_J and S_T stands for Jim's and Tom's strategy and -P and -A show that they have one less move.

<u>4</u>;4	3; <u>5</u>
<u>4</u>;<u>2</u>	<u>4</u>;<u>2</u>

The Nash Equilibria of the second subgame are: {(-A, -A); (-A, -P)}. Let's compare these results with the original game's Nash Equilibria:

The Nash Equilibria of the original game: {**(AA, AA)**; (AP, AP); **(AA, AP)**; (AP, AA)}

We can see that again, the same two equilibria match in subgame perfection with the original game.

Finally, let's analyze the last subgame, where we have only one player, Tom.

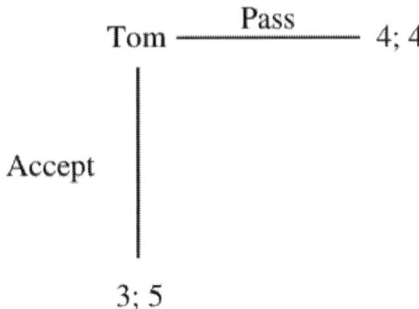

Picture 29: The last subgame.

Tom only has one round with the following strategy profile: S_T= {-P; -A}. His options in normal form look like this:

4;4	3;5

And Tom's best response in the above scenario is 5 (as it is better than 4). There is only one Nash Equilibrium here, namely the {(-A)} strategy. Let's compare it to the main game's Nash Equilibria:

The Nash Equilibria of the original game: {**(AA, AA)**; ~~(AP, AP)~~; ~~(AA, AP)~~; ~~(AP, AA)~~}

This final subgame only matches one Nash Equilibrium in the last game. So, the only equilibrium of the entire game that is also an equilibrium for every subgame is the Accept-Accept, Accept-Accept (AA, AA) subgame perfect equilibrium. We could extend this result to the entire centipede game. The conclusion is that it's always better for all players to play Accept at all times and rounds, which will kill the game off in the first round.

The Problem

The problem with the Centipede - or the Quadruped Game – is that the conclusion is not realistic. Strictly following the principles of game theory, the best response Jim can give is playing Accept right off the bat. But by doing so, Jim will only get two candies. And Tom getting no candies is a shallow consolation when Jim could have had 99 or 100 candies! In real life, Jim and Tom would play Pass for a while until Tom decides to Accept – either in the last round or somewhere along the game.

So why are we getting one result in game theory when we know the outcome would be different in real life?

The issue is not with game theory but how this game is primed. We assume that Tom and Jim are rational players, and their only objective is to maximize their gains over the other player. If we approach the Centipede Game from this angle, the result, 2 for Jim and 0 for Tom, actually makes sense. Using backward induction, we can see how each player would choose Accept, screwing the other player's chances, back to the first move.

Players in real life have multiple motivations to approach a game, maximizing utility (money or candies) is just one of them. What primes the game theory calculation is the story we create for the game. The order of things happening is this: Story →Game Theory → Results, and not Game Theory →Story → Results.[lxi] Game theory is just math we apply to justify or calculate our story's parameters. If we don't like the outcome we got, it's not Game Theory's fault, it's the story's fault. If the conclusion in the Centipede Game is not realistic is because our story wasn't realistic. Players are driven by many other motivations besides maximizing utility.

Let's tell a different story. Let's say our players are Jim and Kat, and they are lovers. And their payoffs are not candies but kisses. They love each other a lot, so they would love the other player to receive more kisses and adoration than they do. Here, the obvious choice would be for both to Pass until the very end of the game, and then both get 100 kisses.

The bottom line is, whenever the outcome of backward induction or any other game theory model doesn't make sense, look for the flaw in the game's story setup, not the game theory model.

This chapter concludes the introduction of extensive form games and this book. I hope you had a good time reading this book, learned things you didn't know before, and can use the information provided in this book to make more strategic choices in your life.

Good luck!

A.R.

Solution for the Health Care Problems Exercise

In Chapter 4, I gave you an exercise to run the mixed-strategy algorithm and then calculate the payoffs for the Health Care Problem. I hope you did it. To compare your results with mine, read this chapter.

Run the Mixed-Strategy Algorithm

Let's start with Jim's mixed strategy. We want to create a mixed strategy that makes ER MD's *expected utility* for selecting quick(q) as a pure strategy, equal to his expected utility (or payoff) if he chooses thorough (t) as a pure strategy. We will mark expected utility as *EU*, quick as $_Q$, and thorough as $_T$. And we will illustrate everything we talked about in this paragraph as:

$$EU_Q = EU_T$$

Remember that mixed strategies are based on a probability. Each part of the equation is a function of a mixed strategy, and represented in the following way:

$$EU_Q = f(\sigma_{ER})$$
$$EU_T = f(\sigma_{ER})$$

In these two equations, f represents the function and σ_{ER} (σ = sigma, $_{ER}$ = Choosing ER) shows the probability that Jim goes to the ER. Wow, we have three equations with three unknowns!

If ER MD goes quick, her payoff depends on the decision Jim makes—whether he will go to the ER or a specialist. If Jim goes to the ER, ER MD gets 5. If Jim goes to the specialist, ER MD gets 0. This is what $EU_Q = f(\sigma_{ER})$ shows.

Jim ↓ / ER MD →	Quick
ER	3; 5
Specialist	4; 0

The same stands true if ER MD decides to do a thorough exam. If Jim goes to the ER, ER MD gets 4. If Jim goes to the specialist, ER MD gets 0. This is what this equation says: $EU_T = f(\sigma_{ER})$.

To solve for $EU_Q = f(\sigma_{ER})$, we need to look at what Jim's expected utility for going quick is as a function of the mixed strategy σ_{ER}. The ER MD will get 5 for a percentage of the time and 0 for some percentage of the time. The equation for the expected utility if ER MD chooses to go quick looks like this:

$$EU_Q = \sigma_{ER}(5) + (1 - \sigma_{ER})(0)$$

To solve for ER MD selecting quick, sigma ER is multiplied by 5 and added to 1 minus sigma ER that has been multiplied by 0. Let's zoom in on each element of the equation above, one by one. Sigma ER (σ_{ER}) shows the probability of Jim going to the ER. That percentage of the time, ER MD's outcome will be 5. We need to add this outcome to what happens the rest of the time. $1 - \sigma_{ER}$ is the probability that Jim goes to a specialist. In this case, ER MD is getting 0 as a payoff. So, we multiply the percentage of the time when Jim goes to the Specialist ($1-\sigma_U$) with ER MD's payoff: 0.

Jim ↓ / ER MD →	Thorough
ER	5; 4
Specialist	4; 0

We have to apply the same logic when ER MD chooses to do a thorough exam. Let's write the equation for the expected utility of ER MD being thorough:

$$EU_T = \sigma_{ER}(4) + (1 - \sigma_{ER})(0)$$

What does this mean? This is the expected utility of being thorough as a function of sigma ER (σ_{ER}). Again, Jim goes to the ER in some cases and then ER MD's payoff is 4. On occasions, Jim will go to a specialist, and then ER MD's payoff is 0. As we see in the equation, sigma ER (σ_{ER}) shows the probability of Jim going to the ER. That percentage of the time, ER MD's outcome will be 4. We need to add this outcome to what happens the rest of the time. $1 - \sigma_{ER}$ is the probability that Jim goes to the specialist. In this case, ER MD is getting 0 as a payoff. So, we multiply the percentage of the time when Jim goes to the specialist ($1-\sigma_{ER}$) with ER MD's payoff: 0.

Now, we know from the first equation that $EU_Q = EU_T$. Let's set both our equations up this way and work out the algebra:

$$\sigma_{ER}(5) + (1 - \sigma_{ER})(0) = \sigma_{ER}(4) + (1 - \sigma_{ER})(0)$$

Let's solve this!

Step 1: Simplify both sides of the equation.
$\sigma_{ER}(5)+(1-\sigma_{ER})(0) = \sigma_{ER}(4)+(1-\sigma_{ER})(0)$
$5\sigma_{ER}+0 = 4\sigma_{ER}+0$
$(5\sigma_{ER})+(0) = (4\sigma_{ER})+(0)$
$5\sigma_{ER} = 4\sigma_{ER}$
$5\sigma_{ER} = 4\sigma_{ER}$

Step 2: Subtract $4\sigma_{ER}$ from both sides.
$5\sigma_{ER}-4\sigma_{ER} = 4\sigma_{ER}-4\sigma_{ER}$
$\sigma_{ER} = 0$

Kaboom! This tells us that in no situation (0 cases) will the ER MD be indifferent between Jim going to the ER vs. going to a specialist. It makes sense, and the doctor gets nothing if Jim chooses to go to a specialist.

Now let's analyze Jim's expected utility for going to the ER or a specialist when those two things are equal? Let's illustrate it like this:

$EU_{ER} = EU_S$

The expected utility of going to the ER is a function of ER MD's quick choice.

$$EU_{ER} = f(\sigma_Q)$$

Similarly, the expected utility of going to a specialist is a function of ER MD's decision of being quick.

$$EU_S = f(\sigma_Q)$$

Let's bring everything together:

Jim ↓ / ER MD →	Quick	Thorough
ER	3; 5	5; 4

Jim's expected utility to go to the ER is 3 if ER MD moves quick and 5 if ER MD is thorough. See the equation below:

$$EU_{ER} = \sigma_Q(3) + (1 - \sigma_Q)(5)$$

Jim ↓ / ER MD →	Quick	Thorough
Specialist	4; 0	4; 0

Jim's expected utility to go to a specialist is 4 if ER MD would move quickly and also 4 if ER MD would be thorough.

$$EU_S = \sigma_Q(4) + (1-\sigma_Q)(4)$$

As $EU_{ER} = EU_S$, we can write our equation as follows:

$$\sigma_Q(3) + (1-\sigma_Q)(5) = \sigma_Q(4) + (1-\sigma_Q)(4)$$

Let's solve this equation step-by-step.

Step 1: Simplify both sides of the equation.
$\sigma_Q(3)+(1-\sigma_Q)(5)= \sigma_Q(4)+(1-\sigma_Q)(4)$
$\sigma_Q(3)+(1)(5)+(-\sigma_Q)(5)= \sigma_Q(4)+(1)(4)+(-\sigma_Q)(4)$
$3\sigma_Q+5+-5\sigma_Q=4\sigma_Q+4+-4\sigma_Q$
$(3\sigma_Q+-5\sigma_Q)+(5)=(4\sigma_Q+-4\sigma_Q)+(4)$
$-2\sigma_Q+5=4$
$-2\sigma_Q+5=4$

Step 2: Subtract 5 from both sides.
$-2\sigma_Q+5-5=4-5$
$-2\sigma_Q=-1$

Step 3: Divide both sides by -2.

$-2^{\sigma}Q/-2 = -1/-2$

$\sigma_Q = \underline{1/2}$

Let's fill our game matrix with all the information we got.

Jim ↓ / ER MD →	Quick 1/2	Thorough 1/2
ER 0	3; 5	5; 4
Specialist 0	4; 0	4; 0

So what does this mean? It is confusing as we haven't encounter mixed-strategy games with a 0 probability. Before I elaborate on what happens in such games, let's calculate the payoffs, just as a practice.

We don't know the actual payoffs of this mixed strategy. What we know are the probabilities.

To make our calculation, we will need to use the following steps.

1. First, we need to find the probability that each outcome happens in equilibrium.
2. Then, we need to multiply that probability with a given player's payoff for each outcome.

3. Sum all these numbers together.[lxii]

We've already found the probabilities of the players. The next step is to multiply across the matrix. To see what the probability of Jim going to the ER and ER MD going quick is, we take Jim's probability, 0, and multiply it with ER MD's probability, 1/2.
0 x 1/2 = 0. When we do the same calculation for each scenario, we realize that they all will be 0. The next step is to take the probabilities and multiply them with their outcomes – breaking news, it all will be 0.

If we'd want to illustrate this outcome, this is how the game tree would look like:

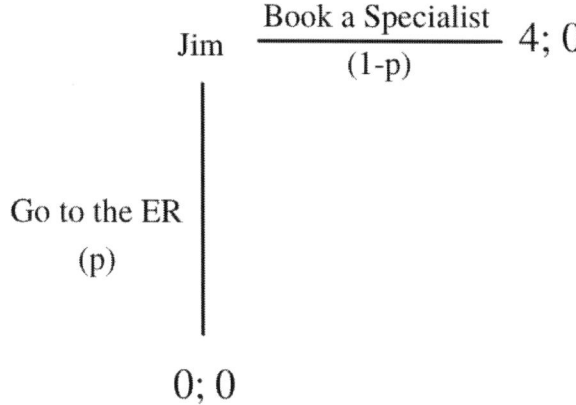

Picture 30: Jim's payoffs if he goes to the ER vs. booking a specialist.

In game theory terms, if we look just at the first step of the game, when Jim goes to the ER with a probability of p, his payoff will be 0. If he goes to a specialist with a probability of 1-p, his payoff will be 4. Considering that the probability we calculated was 0, it is evident that Jim should choose the specialist over the ER.

The ER doctor can't make his moves with any probability as Jim won't go to the ER. The calculation of the game proves that as no credible commitment can happen, there is no chance that Jim would trust the ER doctor. Considering that Jim is going to a specialist, and assuming the ER doctor could make a move, she would be indifferent between performing a quick or thorough exam as her payoff would be 0 either way. So that's our strange mixed-strategy Nash Equilibrium.

Normally, setting the probability of a move to 0 would mean other moves dominate that move. Going to the ER being dominated by going to the specialist means that going to the ER is worse than going to the specialist independent of the other player's, in our case, the ER doctor's, strategy. The move with zero probability would need to be a worse response to the other player's equilibrium strategy than the move with a non-zero probability. As we saw here, if the ER doc could commit to a thorough exam, both players would fare better than with the "going to the specialist" strategy. So there are better responses within the game than going to the specialist. We just never get to that point as, without a credible assurance, Jim would not choose the "go to the ER" move. But what if a credible assurance was possible?

What if the government stepped in as a mediator hypothetically and forced ER doctors to perform a thorough examination? How would this game change?

These are questions to spice up your mind but outside of the scope of this book. In a future book, we will learn how to strategize with mediation involved.

Before You Go…

I would be so very grateful if you would take a few seconds and rate or review this book! Reviews – testimonials of your experience - are critical to an author's livelihood. While reviews are surprisingly hard to come by, they provide the life blood for me being able to stay in business and dedicate myself to the thing I love the most, writing.

If this book helped, touched, or spoke to you in any way, please leave me a review and give me your honest feedback.

CLICK HERE TO REVIEW

Thank you so much for reading this book!

Reference

Binmore, Kenneth (2007). Playing for real: a text on game theory. Oxford University Press US. ISBN 978-0-19-530057-4.

Chandrasekaran, R. (n.d.). Cooperative Game Theory. The University of Texas. Retrieved March 25, 2021, https://personal.utdallas.edu/~chandra/documents/6311/coopgames.pdf

Fang, C., Kimbrough, S. O., Pace, S., Valluri, A., & Zheng, Z. (2002). On Adaptive Emergence of Trust Behavior in the Game of Stag Hunt. Group Decision and Negotiation, 11(6), 449–467. https://doi.org/10.1023/A:1020639132471

Gallego, L. (2017, March 1). *Subgame*. Policonomics. Retrieved January 25, 2022, from https://policonomics.com/subgame-equilibrium/

Harsanyi, John C. (1961). "On the Rationality Postulates underlying the Theory of Cooperative Games". *The Journal of Conflict Resolution.* 5 (2): 179–196. doi:10.1177/002200276100500205. S2CID 220642229.

Jin, X., & Das, S. K. (2012). *Nash Equilibrium - an overview | ScienceDirect Topics.* Handbook on Securing Cyber-Physical Critical Infrastructure. Retrieved January 25, 2022, from https://www.sciencedirect.com/topics/computer-science/nash-equilibrium

Knight, V. (2014). *Chapter 7 - Extensive form games and backwards induction.* VKnight. Retrieved January 25, 2022, from https://vknight.org/Year_3_game_theory_course/Content/Chapter_07-Extensive_form_games_and_backwards_induction/

Knight, V. [Vincent Knight]. (2015, February 17). *Backwards induction and subgame perfection in the Centipede game*[Video]. YouTube. https://www.youtube.com/watch?v=ELLp_dxP_1U

MIT. (2012). *Subgame-Perfect Nash Equilibrium* [Slides]. Ocw.Mit.Edu. https://ocw.mit.edu/courses/economics/14-12-economic-applications-of-game-theory-fall-2012/lecture-notes/MIT14_12F12_chapter11.pdf

Mycielski, Jan (1992). "Games with Perfect Information". Handbook of Game Theory with Economic Applications. 1. pp. 41–70. doi:10.1016/S1574-0005(05)80006-2. ISBN 978-0-4448-8098-7.

Myerson, Roger B. (1991). Game Theory: Analysis of Conflict, Harvard University Press, p. 1. Chapter-preview links, pp. vii–xi.

Osborne, Martin J.; Rubinstein, Ariel (12 Jul 1994). A Course in Game Theory. Cambridge, MA: MIT. p. 14. ISBN 9780262150415.

Osborne, M. J. (2004). *An Introduction to Game Theory*. Oxford University Press.

Owen, Guillermo (1995). *Game Theory: Third Edition*. Bingley: Emerald Group Publishing. p. 11. ISBN 978-0-12-531151-9.

Ozdaglar, A. (2010, March 16). *Game Theory with Engineering Applications Lecture 12: Extensive Form Games* [Slides]. MIT. https://ocw.mit.edu/courses/electrical-engineering-and-computer-science/6-254-game-theory-with-engineering-applications-spring-2010/lecture-notes/MIT6_254S10_lec12.pdf

Özyurt, S. (n.d.). *Game Theory*. Mysite. Retrieved January 25, 2022, from https://www.selcukozyurt.com/game-theory

Picardo, E. (2019, May 19). How Game Theory Strategy Improves Decision Making. Investopedia. https://www.investopedia.com/articles/investing/1

11113/advanced-game-theory-strategies-decisionmaking.asp

Prisner, E. (2014). Game Theory through Examples [eBook edition]. Classroom Resource Materials. https://www.maa.org/sites/default/files/pdf/ebooks/GTE_sample.pdf.

Rasmusen, Eric (2007). Games and Information (4th ed.). ISBN 978-1-4051-3666-2.

Rutherford, A. (2021). *Learn Game Theory: A Primer to Strategic Thinking and Advanced Decision-Making (Strategic Thinking Skills, Book 1)* (1st ed.). Albert Rutherford.

ScienceDirect. (2014). Ultimatum Game - an overview. Retrieved March 24, 2021, from https://www.sciencedirect.com/topics/neuroscience/ultimatum-game

Shoham, Yoav; Leyton-Brown, Kevin (15 December 2008). *Multiagent Systems: Algorithmic, Game-Theoretic, and Logical Foundations*. Cambridge University Press. ISBN 978-1-139-47524-2.

Shor, M. (2005c, August 15). Non-Cooperative Game. Gametheory.net. https://www.gametheory.net/dictionary/Non-CooperativeGame.html

Spaniel, W. (n.d.-a). *Backward Induction – Game Theory 101*. Game Theory 101. Retrieved January 25, 2022, from http://gametheory101.com/courses/game-theory-101/backward-induction/

Spaniel, W. (n.d.-b). *Games with Stages – Game Theory 101*. Game Theory 101. Retrieved January 25, 2022, from http://gametheory101.com/courses/game-theory-101/games-with-stages/

Spaniel, W. (n.d.-c). *Multiple Subgame Perfect Equilibria – Game Theory 101*. Game Theory 101. Retrieved January 25, 2022, from http://gametheory101.com/courses/game-theory-101/multiple-subgame-perfect-equilibria/

Spaniel, W. (n.d.-d). *Subgame Perfect Equilibrium – Game Theory 101.* Game Theory 101. Retrieved January 25, 2022, from http://gametheory101.com/courses/game-theory-101/subgame-perfect-equilibrium/

University of British Columbia. (2007, February 27). *Extensive Form Games and Backward Induction* [Slides]. Www.Cs.Ubc.Ca. https://www.cs.ubc.ca/~kevinlb/teaching/isci330%20-%202006-7/Lectures/lect13.pdf

van Damme, E. (2001). *Extensive Form Game - an overview | ScienceDirect Topics.* Game Theory: Noncooperative Games. Retrieved January 25, 2022, from https://www.sciencedirect.com/topics/computer-science/extensive-form-game

Williams, Paul D. (2013). Security Studies: an Introduction (second ed.). Abingdon: Routledge. pp. 55–56.

Endnotes

[i] Myerson, Roger B. (1991). Game Theory: Analysis of Conflict, Harvard University Press, p. 1. Chapter-preview links, pp. vii–xi.
[ii] Picardo, E. (2019, May 19). How Game Theory Strategy Improves Decision Making. Investopedia. https://www.investopedia.com/articles/investing/111113/advanced-game-theory-strategies-decisionmaking.asp
[iii] Rasmusen, Eric (2007). Games and Information (4th ed.). ISBN 978-1-4051-3666-2.
[iv] Shoham, Yoav; Leyton-Brown, Kevin (15 December 2008). *Multiagent Systems: Algorithmic, Game-Theoretic, and Logical Foundations*. Cambridge University Press. ISBN 978-1-139-47524-2.
[v] Williams, Paul D. (2013). Security Studies: an Introduction (second ed.). Abingdon: Routledge. pp. 55–56.
[vi] Mycielski, Jan (1992). "Games with Perfect Information". Handbook of Game Theory with Economic Applications. 1. pp. 41–70. doi:10.1016/S1574-0005(05)80006-2. ISBN 978-0-4448-8098-7.
[vii] Prisner, E. (2014). Game Theory through Examples [eBook edition]. Classroom Resource Materials. https://www.maa.org/sites/default/files/pdf/ebooks/GTE_sample.pdf.
[viii] Binmore, Kenneth (2007). Playing for real: a text on game theory. Oxford University Press US. ISBN 978-0-19-530057-4.
[ix] Chandrasekaran, R. (n.d.). Cooperative Game Theory.

The University of Texas. Retrieved March 25, 2021, https://personal.utdallas.edu/~chandra/documents/6311/coopgames.pdf

[x] Shor, M. (2005c, August 15). Non-Cooperative Game. Gametheory.net. https://www.gametheory.net/dictionary/Non-CooperativeGame.html

[xi] Shor, M. (2005b, August 12). Symmetric Game. Gametheory.net. https://www.gametheory.net/dictionary/Games/SymmetricGame.html

[xii] Harsanyi, John C. (1961). "On the Rationality Postulates underlying the Theory of Cooperative Games". The Journal of Conflict Resolution. 5 (2): 179–196. doi:10.1177/002200276100500205. S2CID 220642229.

[xiii] ScienceDirect. (2014). Ultimatum Game - an overview. Retrieved March 24, 2021, from https://www.sciencedirect.com/topics/neuroscience/ultimatum-game

[xiv] Owen, Guillermo (1995). Game Theory: Third Edition. Bingley: Emerald Group Publishing. p. 11. ISBN 978-0-12-531151-9.

[xv] Jin, X., & Das, S. K. (2012). Nash Equilibrium - an overview | ScienceDirect Topics. Handbook on Securing Cyber-Physical Critical Infrastructure. Retrieved January 25, 2022, from https://www.sciencedirect.com/topics/computer-science/nash-equilibrium

[xvi] Rutherford, A. (2021). *Learn Game Theory: A Primer to Strategic Thinking and Advanced Decision-Making (Strategic Thinking Skills, Book 1)* (1st ed.). Albert Rutherford.

[xvii] Rutherford, A. (2021). Learn Game Theory: A Primer to Strategic Thinking and Advanced Decision-Making

(Strategic Thinking Skills, Book 1) (1st ed.). Albert Rutherford.
[xviii] Spaniel, W. (2011). Game Theory 101: The Complete Textbook. Self-published.
[xix] Fang, C., Kimbrough, S. O., Pace, S., Valluri, A., & Zheng, Z. (2002). On Adaptive Emergence of Trust Behavior in the Game of Stag Hunt. Group Decision and Negotiation, 11(6), 449–467. https://doi.org/10.1023/A:1020639132471
[xx] Spaniel, W. (2011). Game Theory 101: The Complete Textbook. Self-published.
[xxi] Rutherford, A. (2021). Learn Game Theory: A Primer to Strategic Thinking and Advanced Decision-Making (Strategic Thinking Skills, Book 1) (1st ed.). Albert Rutherford.
[xxii] Osborne, Martin J.; Rubinstein, Ariel (12 Jul 1994). A Course in Game Theory. Cambridge, MA: MIT. p. 14. ISBN 9780262150415.
[xxiii] Jin, X., & Das, S. K. (2012). Nash Equilibrium - an overview | ScienceDirect Topics. Handbook on Securing Cyber-Physical Critical Infrastructure. Retrieved January 25, 2022, from https://www.sciencedirect.com/topics/computer-science/nash-equilibrium
[xxiv] Spaniel, W. (2011). Game Theory 101: The Complete Textbook. Self-published.
[xxv] Spaniel, W. (2011). Game Theory 101: The Complete Textbook. Self-published.
[xxvi] Rutherford, A. (2021). Learn Game Theory: A Primer to Strategic Thinking and Advanced Decision-Making (Strategic Thinking Skills, Book 1) (1st ed.). Albert Rutherford.
[xxvii] Rutherford, A. (2021). Learn Game Theory: A Primer to Strategic Thinking and Advanced Decision-Making

(Strategic Thinking Skills, Book 1) (1st ed.). Albert Rutherford.

xxviii Ozdaglar, A. (2010, March 16). Game Theory with Engineering Applications Lecture 12: Extensive Form Games [Slides]. MIT. https://ocw.mit.edu/courses/electrical-engineering-and-computer-science/6-254-game-theory-with-engineering-applications-spring-2010/lecture-notes/MIT6_254S10_lec12.pdf

xxix MIT. (2012). Subgame-Perfect Nash Equilibrium [Slides]. Ocw.Mit.Edu. https://ocw.mit.edu/courses/economics/14-12-economic-applications-of-game-theory-fall-2012/lecture-notes/MIT14_12F12_chapter11.pdf

xxx Spaniel, W. (n.d.). Subgame Perfect Equilibrium – Game Theory 101. Game Theory 101. Retrieved January 25, 2022, from http://gametheory101.com/courses/game-theory-101/subgame-perfect-equilibrium/

xxxi Osborne, M. J. (2004). An Introduction to Game Theory. Oxford University Press.

xxxii Spaniel, W. (n.d.). Subgame Perfect Equilibrium – Game Theory 101. Game Theory 101. Retrieved January 25, 2022, from http://gametheory101.com/courses/game-theory-101/subgame-perfect-equilibrium/

xxxiii Spaniel, W. (n.d.). Subgame Perfect Equilibrium – Game Theory 101. Game Theory 101. Retrieved January 25, 2022, from http://gametheory101.com/courses/game-theory-101/subgame-perfect-equilibrium/

xxxiv van Damme, E. (2001). Extensive Form Game - an overview | ScienceDirect Topics. Game Theory: Noncooperative Games. Retrieved January 25, 2022, from https://www.sciencedirect.com/topics/computer-science/extensive-form-game

xxxv Knight, V. (2014). Chapter 7 - Extensive form

games and backwards induction. VKnight. Retrieved January 25, 2022, from https://vknight.org/Year_3_game_theory_course/Content/Chapter_07-Extensive_form_games_and_backwards_induction/
[xxxvi] Ozdaglar, A. (2010, March 16). Game Theory with Engineering Applications Lecture 12: Extensive Form Games [Slides]. MIT. https://ocw.mit.edu/courses/electrical-engineering-and-computer-science/6-254-game-theory-with-engineering-applications-spring-2010/lecture-notes/MIT6_254S10_lec12.pdf
[xxxvii] Knight, V. (2014). Chapter 7 - Extensive form games and backwards induction. VKnight. Retrieved January 25, 2022, from https://vknight.org/Year_3_game_theory_course/Content/Chapter_07-Extensive_form_games_and_backwards_induction/
[xxxviii] Spaniel, W. (n.d.-a). Backward Induction – Game Theory 101. Game Theory 101. Retrieved January 25, 2022, from http://gametheory101.com/courses/game-theory-101/backward-induction/
[xxxix] Spaniel, W. (n.d.-a). Backward Induction – Game Theory 101. Game Theory 101. Retrieved January 25, 2022, from http://gametheory101.com/courses/game-theory-101/backward-induction/
[xl] Gallego, L. (2017, March 1). Subgame. Policonomics. Retrieved January 25, 2022, from https://policonomics.com/subgame-equilibrium/
[xli] Gallego, L. (2017, March 1). Subgame. Policonomics. Retrieved January 25, 2022, from https://policonomics.com/subgame-equilibrium/
[xlii] Rutherford, A. (2021). Learn Game Theory: A Primer to Strategic Thinking and Advanced Decision-Making

(Strategic Thinking Skills, Book 1) (1st ed.). Albert Rutherford.
[xliii] Spaniel, W. (2011). Game Theory 101: The Complete Textbook. Self-published.
[xliv] Rutherford, A. (2021). Learn Game Theory: A Primer to Strategic Thinking and Advanced Decision-Making (Strategic Thinking Skills, Book 1) (1st ed.). Albert Rutherford.
[xlv] Spaniel, W. (2011). Game Theory 101: The Complete Textbook. Self-published.
[xlvi] Spaniel, W. (n.d.-b). Multiple Subgame Perfect Equilibria – Game Theory 101. Game Theory 101. Retrieved January 25, 2022, from http://gametheory101.com/courses/game-theory-101/multiple-subgame-perfect-equilibria/
[xlvii] Spaniel, W. (2011). Game Theory 101: The Complete Textbook. Self-published.
[xlviii] Spaniel, W. (2011). Game Theory 101: The Complete Textbook. Self-published.
[xlix] Spaniel, W. (2011). Game Theory 101: The Complete Textbook. Self-published.
[l] Spaniel, W. (n.d.-b). Games with Stages – Game Theory 101. Game Theory 101. Retrieved January 25, 2022, from http://gametheory101.com/courses/game-theory-101/games-with-stages/
[li] Spaniel, W. (n.d.-b). Games with Stages – Game Theory 101. Game Theory 101. Retrieved January 25, 2022, from http://gametheory101.com/courses/game-theory-101/games-with-stages/
[lii] Spaniel, W. (n.d.-b). Games with Stages – Game Theory 101. Game Theory 101. Retrieved January 25, 2022, from http://gametheory101.com/courses/game-theory-101/games-with-stages/
[liii] Spaniel, W. (n.d.-b). Games with Stages – Game

Theory 101. Game Theory 101. Retrieved January 25, 2022, from http://gametheory101.com/courses/game-theory-101/games-with-stages/

[liv] Rutherford, A. (2021). Learn Game Theory: A Primer to Strategic Thinking and Advanced Decision-Making (Strategic Thinking Skills, Book 1) (1st ed.). Albert Rutherford.

[lv] Özyurt, S. (n.d.). Game Theory. Mysite. Retrieved January 25, 2022, from https://www.selcukozyurt.com/game-theory

[lvi] University of British Columbia. (2007, February 27). Extensive Form Games and Backward Induction [Slides]. Www.Cs.Ubc.Ca. https://www.cs.ubc.ca/~kevinlb/teaching/isci330%20-%202006-7/Lectures/lect13.pdf

[lvii] Spaniel, W. (2011). Game Theory 101: The Complete Textbook. Self-published.

[lviii] University of British Columbia. (2007, February 27). Extensive Form Games and Backward Induction [Slides]. Www.Cs.Ubc.Ca. https://www.cs.ubc.ca/~kevinlb/teaching/isci330%20-%202006-7/Lectures/lect13.pdf

[lix] Knight, V. [Vincent Knight]. (2015, February 17). Backwards induction and subgame perfection in the Centipede game [Video]. YouTube. https://www.youtube.com/watch?v=ELLp_dxP_lU

[lx] Knight, V. [Vincent Knight]. (2015, February 17). Backwards induction and subgame perfection in the Centipede game [Video]. YouTube. https://www.youtube.com/watch?v=ELLp_dxP_lU

[lxi] Spaniel, W. (2011). Game Theory 101: The Complete Textbook. Self-published.

[lxii] Spaniel, W. (2011). Game Theory 101: The Complete Textbook. Self-published.

Printed in Great Britain
by Amazon

67772f88-e135-413a-8ca7-d1c54dc106d8R01